TAKE-HOME PHYSICS:

65 High-Impact, Low-Cost Labs

TAKE-HOME PHYSICS: 65 High-Impact, Low-Cost Labs

Michael Horton

NSTA press
National Science Teachers Association
Arlington, Virgina

Claire Reinburg, Director
Jennifer Horak, Managing Editor
Judy Cusick, Senior Editor
Andrew Cocke, Associate Editor
Wendy Rubin, Associate Editor

ART AND DESIGN
Will Thomas Jr., Director
Tim French, Senior Graphic Designer, cover and interior design

PRINTING AND PRODUCTION
Catherine Lorrain, Director

SCILINKS
Tyson Brown, Director
Virginie L. Chokouanga, Customer Service and Database Coordinator

NATIONAL SCIENCE TEACHERS ASSOCIATION
Francis Q. Eberle, Executive Director
David Beacom, Publisher

12 11 10 09 4 3 2 1

LIBRARY OF CONGRESS CATALOGING-IN-PUBLICATION DATA
Horton, Michael, 1972-
Take-home physics : 65 high-impact, low-cost labs / by Michael Horton.
p. cm.
Includes bibliographical references and index.
ISBN 978-1-935155-05-8
1. Physics--Experiments--Laboratory manuals. 2. Physics--Study and teaching (Secondary)--Activity programs. I. Title.
QC35.H795 2009
530.071'2--dc22

2009013538

NSTA is committed to publishing material that promotes the best in inquiry-based science education. However, conditions of actual use may vary, and the safety procedures and practices described in this book are intended to serve only as a guide. Additional precautionary measures may be required. NSTA and the authors do not warrant or represent that the procedures and practices in this book meet any safety code or standard of federal, state, or local regulations. NSTA and the authors disclaim any liability for personal injury or damage to property arising out of or relating to the use of this book, including any of the recommendations, instructions, or materials contained therein.

Featuring sciLINKS®—a new way of connecting text and the Internet. Up-to-the minute online content, classroom ideas, and other materials are just a click away. For more information go to www.scilinks.org/faq/moreinformation.asp.

TABLE OF CONTENTS

SECTION 2: Forces and Energy

SECTION 3: Waves, Sound, and Light

Section 4: Electricity and Magnetism

INTRODUCTION

Research has shown that homework can be an effective and meaningful learning tool for high school students if it is relevant, engaging, and hands-on. These take-home physics activities are designed to match those criteria. Educational writer Alfie Kohn said in a 2006 interview that there are only two ways that homework is effective for high school students. One of those is "activities that have to be done at home, such as…a science experiment in the kitchen" (Oleck 2006).

This book is a collection of physics labs that lend themselves to being performed at home with simple materials. It is not intended to be a physics textbook or to cover every topic encountered in high school physics. Most of the labs are written as Structured or Level 2 inquiry (see Inquiry in Physics, page xi), and some include instructions to raise the level of inquiry if the teacher feels comfortable doing so. A few activities just aren't compatible with inquiry at home and are written as verification labs. Most of the activities involve measuring, graphing, calculating, extrapolating graphs, and other science-process skills.

Because this is one piece of a complete physics curriculum, it is assumed that traditional learning and hands-on activities in the classroom will fill in where take-home labs are not practical. Teachers may choose to eliminate some of the labs and substitute others without breaking the flow of the labs.

Used in this way, the hands-on activities can be a powerful tool for learning physics concepts and preparing students for physics assessments that are highly dependent on charts, graphs, and conceptual questions. These activities have been piloted in schools across the United States and used by teachers who received the material during conference presentations. The success that these teachers have had with the labs helps refute the common misconception among teachers and students that lectures are for learning and labs are for fun. Students *can* learn physics from labs.

Although the labs are written as take-home activities for high school students, many of the activities in the book are well suited for home-schooled students as well as those who take online courses. These activities would also be appropriate for family science nights and museum outreach programs. If a teacher does not have sufficient materials to send an activity home with every student, the lab could be performed in class as an alternative. One teacher used the activities in after-school intervention programs for students who were not proficient after being exposed to the concepts in the classroom.

Why Take-Home Labs?

These take-home labs, if implemented effectively, can address most or all of the following problems, which are common with physics labs and homework:

Students won't do homework. This sentiment could be rephrased as *Students won't do busywork at home*. When presented with fun and challenging assignments that open doors to understanding the physical world around them, students will rise to the occasion. Throughout four years of implementing these activities, the homework completion rate improved greatly, and test scores indicate that student achievement increased as well.

Students do poorly on standardized tests. Most standardized science tests are weighted toward science-process skills. These skills include drawing and interpreting charts and graphs, finding patterns, interpreting diagrams, and analyzing experimental data. These take-home labs contribute to the improvement of every one of those skills.

There is not enough class time to cover all the standards. A great deal of class time is used on simple labs that students could do at home. These labs are important, but they consume valuable class time. By having students perform these labs at home, teachers recover days of class time in which new concepts can be taught or reinforced.

Students do not experience enough labs. By assigning take-home lab work, teachers can increase the number of labs students complete over the course of the year to nearly 100 labs in class and at home. This will lead to a more engaging, fulfilling learning experience for students, which will lead to deeper, more lasting learning.

Physics labs are expensive. There are 65 labs in this book, and the collection of materials needed to complete the labs costs around $20, or approximately 31¢ per lab per student the first year. After the first year, only breakages and consumables (mostly batteries) have to be replaced at a cost of less than a couple of dollars per kit.

Physics students lack basic skills when they get to my class. Some of the take-home labs teach about background information and skills that students are supposed to remember from middle school but rarely do. These labs are a good refresher that you can refer back to throughout the year. As mentioned earlier, it can also buy you days of class time to teach the physics curriculum for your grade level.

My school will not let me do take-home labs because of No Child Left Behind. NCLB prohibits some schools from allowing take-home labs because the materials are not provided to every student. That practice, it is argued, puts some students at a disadvantage. However, teachers can easily provide almost every object needed to do *Take-Home Physics* labs, with the exception of a couple of labs that require students to provide water or other common materials. And if students tell the teacher in advance that they do not have certain required items at home, the teacher can provide these materials as well.

Evidence of Success

A chemistry teaching colleague and I have collected data on the success of these take-home labs. Most of the physics students were 12th graders, but in California 12th graders do not take state tests, and therefore, it was impossible to study the effectiveness of the labs with an independent physics exam. Instead, the effectiveness of the take-home activities was studied among chemistry students. Although the connection to physics may be anecdotal without a state exam, similar results were seen in exams and in student morale for both chemistry and physics classes.

As shown in Figure 1.1, four years of implementing chemistry take-home labs coincided with a nearly tenfold increase (from 3% to 32%) in the proportion of proficient and advanced students in chemistry based on the 2002 baseline. In the same time period, the number of students testing below basic and far below basic decreased

Figure 1.1

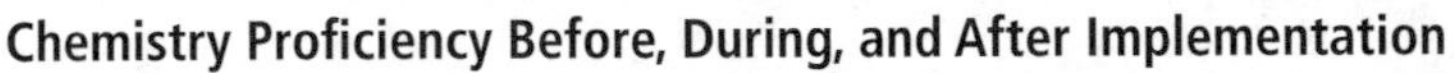

Chemistry Proficiency Before, During, and After Implementation

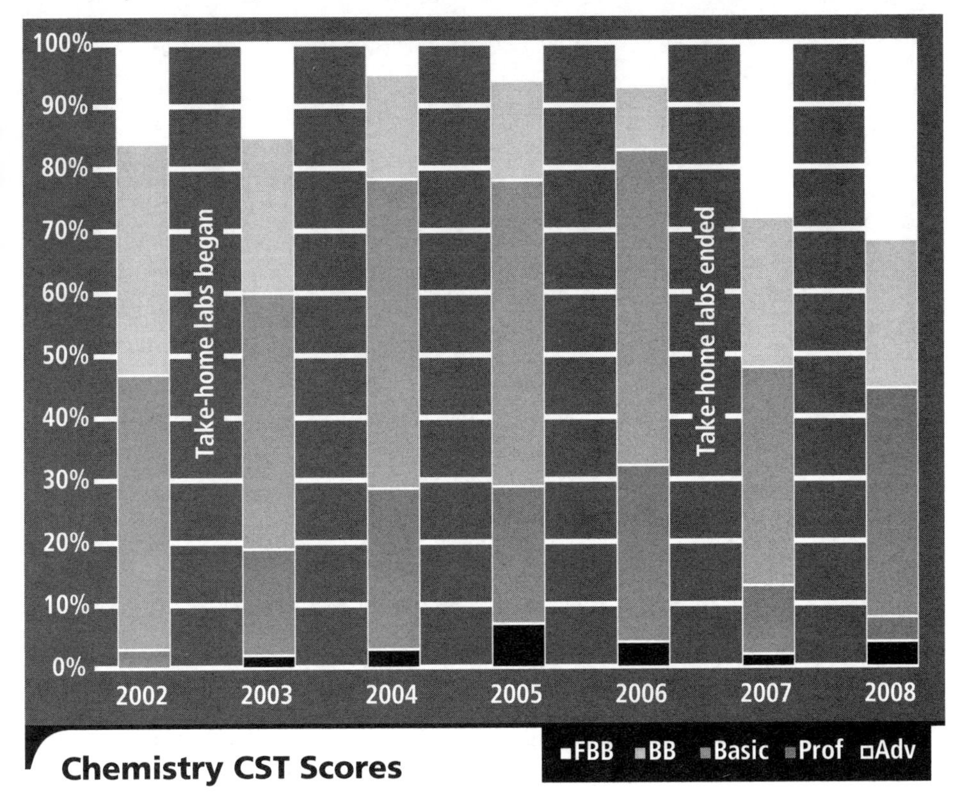

FBB = Far Below Basic, BB = Below Basic, Prof = Proficient, and Adv = Advanced. California considers FBB, BB, and Basic to be nonproficient.

Figure 1.2

Comparison of Student Proficiency in Chemistry, Geometry, and Algebra 2

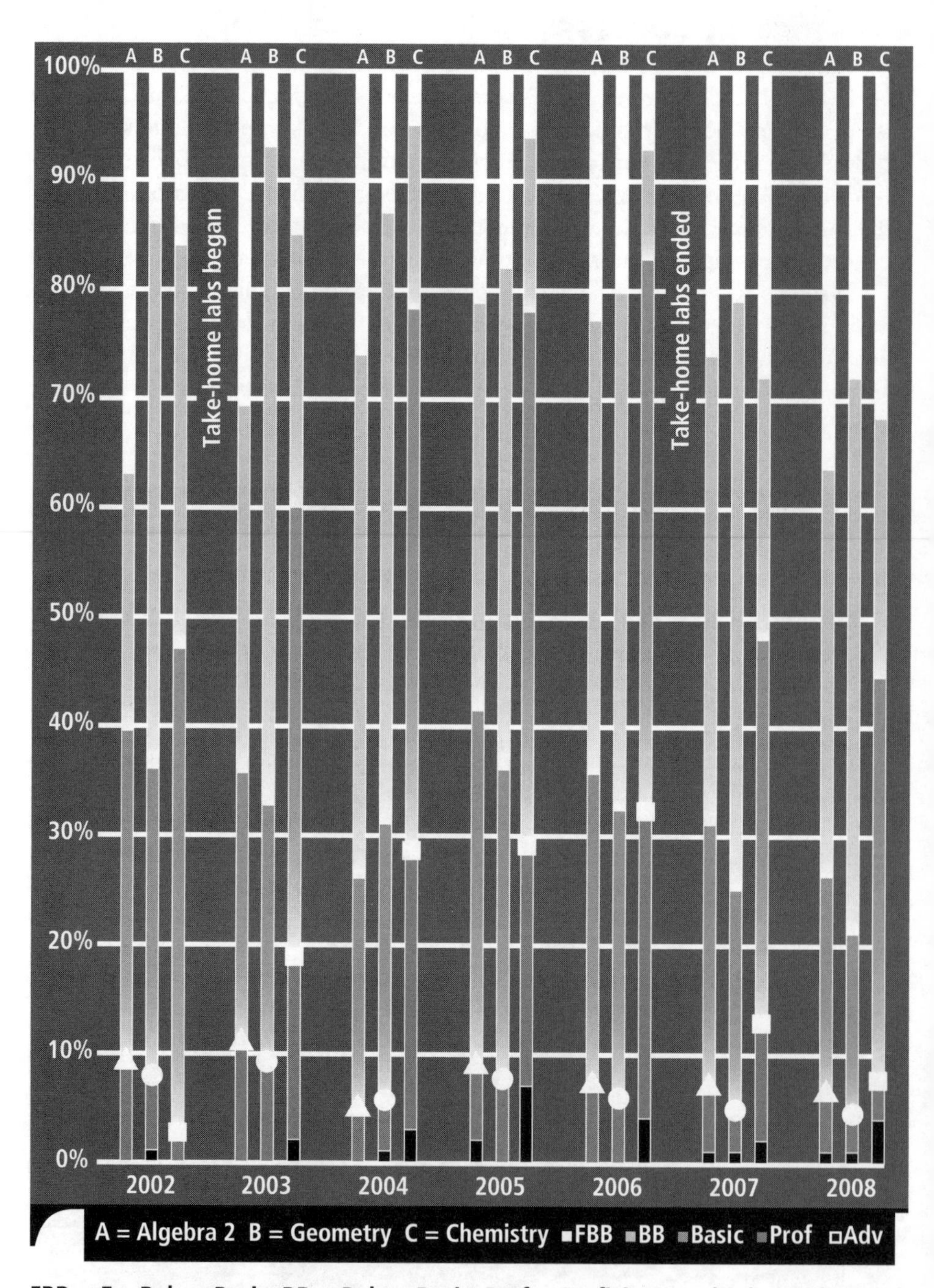

FBB = Far Below Basic, BB = Below Basic, Prof = Proficient, and Adv = Advanced. California considers FBB, BB, and Basic to be nonproficient.

dramatically (53% to 17%). Each year, more activities were added and the procedures were refined.

To strengthen the argument that the score changes were due to the take-home labs and not to other changes in the school or the student body, student performance in Algebra 2 and Geometry over the same period was examined (see Figure 1.2). Those scores stayed nearly the same over this time period. Most chemistry students are in either Algebra 2 or Geometry.

Although I left the classroom at the beginning of the 2005–2006 school year, my chemistry colleague collaborated with the long-term substitute teacher and continued the take-home labs. Figure 1.1 shows that in 2007, when the chemistry colleague became an administrator and the take-home labs ended, the test scores went back almost to where they were before we started. That trend continued in 2008.

My high school was a state-identified underperforming school with more than 70% of the students qualifying for free or reduced-price lunch. More than 40% of parents never graduated from high school and more than 25% of students are English language learners. The state also puts schools into groups of 100 similar schools based on dozens of criteria. My school had far higher chemistry scores than those similar schools, as much as a 156-times-larger ratio of proficient and advanced students to below-basic and far-below-basic students. Overall, the school was ranked in the second decile from the bottom in the state in 2006. In chemistry, the school was above state average, putting it at least five deciles from the bottom. In short, the chemistry scores at this school were above school, district, county, and state averages.

All data in Figures 1.1 and 1.2 were taken from the California Department of Education website (*cde.ca.gov*) for Perris High School in the Perris Union High School District in Riverside County, California.

Inquiry in Physics

The topic of inquiry in science instruction—what it looks like, how best to implement it in the classroom, ways to assess its success—far exceeds the scope of this introduction. The following is a brief explanation of inquiry and how it's defined in this volume. For a full and more detailed discussion of inquiry, please consult the reference list.

When inquiry is discussed in science education, it takes two forms. The first meaning is the creation of an atmosphere of inquiry in the classroom in which students interact with one another and the teacher facilitates open-ended investigation of student-generated questions. The second use of the term *inquiry* refers to inquiry activities. This book alone cannot accomplish the first task, but it is intended to provide the second. The postlab questions take the place of the teacher's guiding questions during an inquiry activity performed in the classroom.

It is a common misconception that inquiry is all or nothing. Most of the research about inquiry activities identifies four levels of inquiry (Banchi and Bell 2008; Colburn 2000; McComas 2005), although the first level is not inquiry at all (see Figure 1.3). What changes at each level is how much information is given to the student (i.e., the question, the procedure, the answer) (Bell, Smetana, and Binns 2005). Herron (1971) first described inquiry by distinguishing among three different levels. Since then, rubrics and matrices have been created with three, four, five, and seven levels (Lee et al. 2001). Because most education researchers refer to four levels, that is the model that will be used here.

Figure 1.3

What Is Provided to Students at the Different Levels of Inquiry

	Question Given	Procedure Given	Answer Known
Level 1	X	X	X
Level 2	X	X	
Level 3	X		
Level 4			

Level 1 inquiry activities are also known as verification, or "cookbook," activities. In a Level 1 activity, students are given the question or problem, they are given the procedure, and they already know the answer to the question. They are simply verifying something that they already learned. This is the least effective form of inquiry.

In a **Level 2,** or "structured inquiry," activity, students are given the question and the procedure, but they do not yet know the answer. Having students do an activity before learning the concept raises the activity to Level 2. In performing this significant step, the teacher is giving all students the same background knowledge, an activity on which to hang the concept in their memory, and a common experience to refer back to in class when covering or practicing the concept. Robert Marzano (2004, p. 3) says, "Students who have a great deal of background knowledge in a given subject area are likely to learn new information readily and quite well. The converse is also true." Dochy, Segers, and Buehl (1999) found through meta-analysis that differences in students' prior knowledge explain 81% of the variance in posttest scores. The research of Langer (1984) and Stevens (1980) shows a well-established correlation between prior knowledge and academic achievement.

Most of the labs in this book are written as Level 2 inquiry, which means the teacher must assign the activities to students before teaching the concept in class. Assigning the activities this way increases the level of inquiry and also allows teachers to afford students prior knowledge when the concept is covered later. Marzano states that "what students *already know* about the content is one of the strongest indicators of how well they will learn new information relative to the content" (2004, p. 1). Douglas Llewellyn, in *Teaching High School Science Through Inquiry: A Case Study Approach* (2005), recommends doing the lab first to raise the inquiry level. Bell (Bell, Smetana, and Binns 2005, p. 33) comments, "The difference between a Level 1 and Level 2 activity can be a matter of timing. A confirmation lab can become a structured inquiry lab by simply presenting the lab before the target concept is taught."

In a **Level 3** activity, referred to as "guided inquiry," students are given an appropriate question and are asked to determine the procedure and develop the answer on their own. Many activities can be converted to Level 3, simply by removing the procedure and having students determine how to accomplish the task. For example, a lab inviting students to follow a procedure to determine what affects the period of a pendulum could be increased to Level 3 by changing the activity to "Which affect(s) the period of a pendulum—length, angle of release, and/or mass of the pendulum bob?" Some of these take-home labs are already Level 3 and teachers are free to modify others to meet the Level 3 criteria. Most of the extension activities, which are often provided at the end of the activities, are opportunities for Level 3 inquiry, and several of the post-lab questions encourage students to test ideas on their own. Teachers can remove some steps in the procedure, add unnecessary steps and have students identify them, remove data tables and have students create their own, add extensions, or rearrange the steps in order to transition to Level 3 gradually (Llewellyn 2005). Some of the labs in this book already expect students to create the data charts.

In a **Level 4** activity, referred to as "open inquiry," students pose their own question and are given the resources to answer that question. This type of inquiry is most easily demonstrated with science fair projects. Students investigate their own questions following their own procedures and draw their own conclusions. Many teachers also use this type of inquiry as differentiation activities. If a student clearly understands a concept while doing an inquiry lab, he or she may be invited to come up with additional questions to answer independently. It is not the intent of this book to provide Level 4 inquiry activities, but teachers are encouraged to motivate students to perform deeper follow-up activities to answer questions that they may have after an activity. Students can certainly extend experiments to answer their own questions, hence independently creating their own Level 4 inquiry activities. Some of the extension activities could be replaced with a more general Level 4 question, such as "What other questions do you have about this topic? Create an activity that will lead you to the answer."

Many teachers are discouraged when they begin using inquiry activities and do not immediately see the achievement gains they expected. Students need experience using inquiry activities to learn. They do not relate classroom lab activities with learning because they have always done verification labs. I have seen low performance the first few times questions were given on a test about a subject learned via inquiry. However, when students were reminded of the activity, they immediately began writing again and performance increased greatly. After a little experience, students no longer need to be reminded and make the connection automatically. Please be warned that providing question stems such as "Remembering the string and protractor lab…" is never a good idea because high-stakes tests will not do so.

Some of the labs in this book are not inquiry-based. They are included to overcome student misconceptions, provide experiences that allow students to see physics phenomena, or enable students to collect and analyze data that will assist them in understanding a concept more thoroughly. Some topics just do not lend themselves well to inquiry at home, but they are far fewer than those topics that do.

Teacher Feedback

Keep in mind that feedback of some type should be given on each lab that students do. This can be time consuming but very valuable.

The purpose of these labs is not to add more items to the teacher's grade book. Homework should be used as formative assessment. Often, formative assessments are not even graded. The teacher uses them to judge where the students are in relation to proficiency on the relevant standard. If the purpose of the lab is to give common background information to the students, then the teacher may also consider this a situation that need not be graded formally. But keep in mind that whether the teacher grades the labs or not, there must be feedback in the classroom. Some teachers may choose to give the labs credit/no credit or credit/redo marks.

Marzano (2006) reports that students who do not receive feedback after a formative assessment do no better on the summative assessment than if they had not been given the formative assessment at all. He lists several different ways of offering that feedback and gives the pre- and posttest gains that can be expected. Just putting a grade at the top of the paper has a negative effect. If the teacher decides to grade the labs, then the feedback should be deeper than a simple overall grade. Having students prepare short written summaries, explain their logic aloud to the class, or discuss the activities while the teacher rotates around to deal with misconceptions are examples of ways teachers can check for understanding without formal grading. It is also helpful for the teacher to identify the criteria used to deem a conclusion satisfactory and allow students to redo the activity until it is satisfactory.

Many schools have not gotten to the point where formative assessments are used this way, however. Teachers may want to give credit for these labs simply to encourage students to perform the labs. Although it is worth repeating that these activities do not have to be graded, following are some tips for easing that burden if the teacher chooses to do so. A compromise could be that the teacher only grades some of the labs but the students do not know in advance which labs will be graded.

Teachers should create a grading guide that identifies which parts of each lab are the most important and assesses only those parts. For those labs that are graded, not every section has to be graded. And for every section that is graded, not every question or detail has to be analyzed.

A grading guide could look something like this:

Name___________________________ Period____________________

Dist/Time Graphs 1
Graphs _____/ 10 Total ____/10

Dist/Time Graphs 2
Data ____ /5 Graphs ____/5 Total ____/10

Average Speed
not graded

Acceleration of G 1
Post-lab Questions ___/10 Total ____/10

Acceleration of G 2
Data ____/5 Conclusion ____/2 Total ____/7

Final Velocity
Data ____/2 Graph ____/5 Total ____/7

Reaction Time not graded

Completeness and Setup of Lab Notebook Total ____/15

Grand Total ____/59

Lab Notebooks

Because students may perform as many as 65 activities, a lab notebook is a useful tool for students to keep track of their lab write-ups; plus, teachers may choose to collect the notebooks after, say, five or ten activities have been completed as a means of assessment. The teacher should determine in advance the format for the notebooks (e.g., Purpose, Procedure, Data, Calculations), and establish a set of rules for using the notebooks. These rules help students stay organized and save teachers time when reviewing the notebooks. The students' write-ups should include data charts, graphs, and post-lab questions, as called for in a given activity's directions. Composition-style notebooks, spiral notebooks, or three-ring binders are all acceptable "notebooks." Teachers may consult a local university instructor to find out how freshmen are required to organize their notebooks, and then ask their own students to set up their notebooks in a similar fashion.

It is clear when students cheat on a take-home lab. For most of the labs, there is a very small chance that any two people would get the exact same answer. If two students do get the exact same answers, it is clear that they copied. They cannot say that they were partners like with an in-class lab as long as the teacher makes it clear that each student is expected to collect his or her own data.

Again, it is not imperative to grade every lab or every part of those that are graded. Formal summative assessments will determine which students did the work and which students did not. These labs are not about assigning grades. They are about learning physics. Each lab should be reviewed by the teacher to check for understanding, and feedback should be given in one form or another.

Tips for Using This Book

Teachers may opt to use *Take-Home Physics* labs exactly as they are presented or adapt them to better suit the needs of a particular class. Labs may be written on the board or posted via overhead projector for students to copy into their lab notebooks, or teachers may choose to pass out copies to each student. Copies, of course, can be stored in materials boxes and reused from year to year.

Figure 1.4

Small Sample of Materials for Lab Boxes

Figure 1.5

Some Items Ready to Distribute

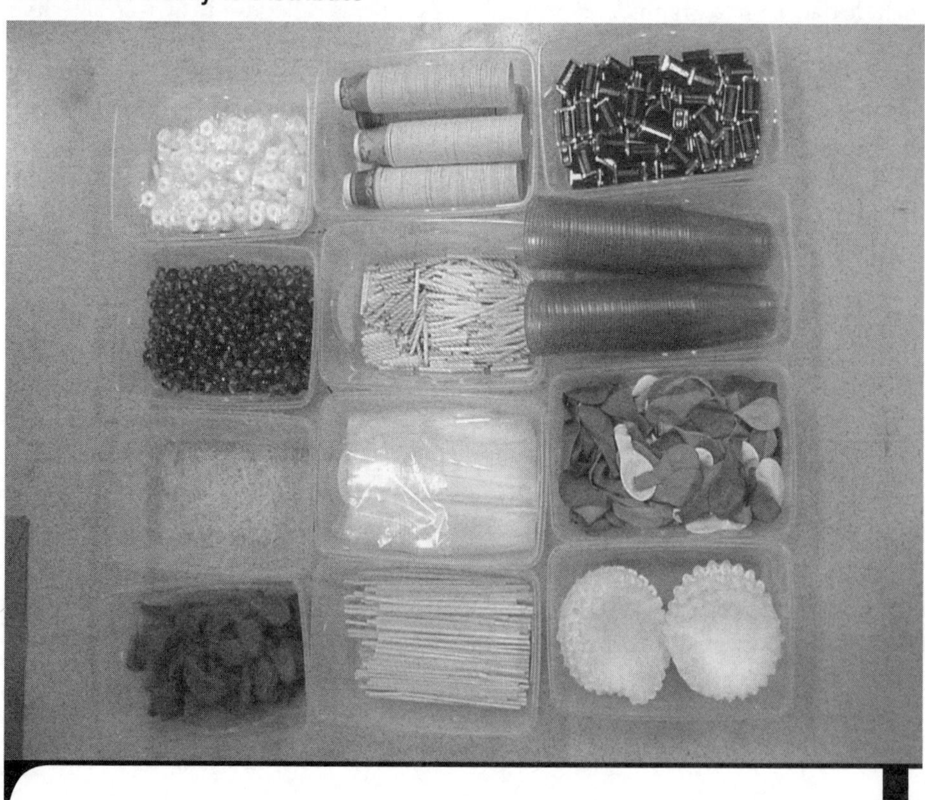

Assembling the Materials

Once you decide to assign the labs in this book, you will need to start planning. A list of materials is included in each lab to assist teachers in determining what materials will be needed. A master materials list is included at the end of this section.

The basic idea is that you will send a (plastic shoe-box size) box of materials home with the students with which they can complete all of the activities in this book. The box is pictured in Figures 1.4 and 1.5 (p. xvi). Sending all of the materials home in one box is preferable to sending individual activities home in plastic sandwich bags because of the time that it will take to check in and out each of the bags. All of the materials can fit into one box, which may be purchased for around a dollar, or alternatively a 1-gallon plastic freezer bag.

After determining which labs you will assign and how much of each item is necessary, it's time to go shopping. Most of the items can be found at discount stores—from the boxes themselves to the cups, batteries, and magnets. A few items such as nails, sandpaper, and washers can be purchased at a home improvement store. Many items can be purchased at a much lower cost online than in stores or catalogs. For example, 12 resistors at a local electronics store go for about $2.50, but on sites like eBay, you may find a roll of 1,500 resistors for $10. The two most expensive materials are the multimeters and the stopwatches. Multimeters can be purchased for around $5 from many electronics distributors online or from hardware stores such as Harbor Freight Tools (*http://harborfreight.com*). Stopwatches can be purchased in larger quantities for about $6 each from education supply companies. Rubber balls can be purchased online very inexpensively, but beware of high shipping costs. The website *www.rebeccas.com* sells 250 bouncy balls for less than $22, including shipping. At an electronics store, two stopwatch batteries can cost $3.75, but *www.cheapbatteries.com* sells 100 for 10¢ each.

Be sure to buy a few extra of each item in case of loss and breakage. Once purchased, the items will take up quite a bit of space. (Figure 1.4 shows a small sample of the materials used to make lab boxes.) Don't plan on storing them in the back of your classroom. (Tip: Purchase a few extra boxes to store items in while you dispense them.)

When all materials have been purchased, put each item in its own box for easier distribution (see Figure 1.5). Find a large space to spread out the boxes in an assembly line so students can walk by, pick up the boxes, and fill them with the materials. Attach a 3 × 5 in. index card with the quantity that students need to take written on it to each box and instruct students to follow the cards in order (for example, pick up six marbles, four paper clips, one rubber ball, etc.).

Remind students that they are to return all nonconsumable items with the boxes at the end of the year or upon transferring to another class or school.

Master Materials List

* Students will be expected to have these items at home. If they do not, the teacher should have some extras to loan out.

- Aluminum foil (20 x 20 cm)
- Batteries, 9 V (2)
- Batteries, D-cell (2)

- Baking soda (50 ml)
- Balloon, helium (optional)
- Balloons, standard (2)
- *Bathroom scale
- Bead, wooden or plastic with a hole through the center, size of a standard marble (can be taken off a beaded necklace from a discount store)
- Bell wire, enameled (~1 m)
- Boat drawings on transparency (see p. 94)
- Bouncy ball, any size
- *Calculator
- *Cans of soup (1 clear, e.g., chicken broth; 1 creamy, e.g., cream of mushroom)
- Card (3 x 5 in.) with hole punched (anywhere on card)
- Cardboard, corrugated (8 ½ x 11 in.)
- Cardboard or thick paper (8 ½ x 11 in.)
- CD case or other piece of hard plastic
- Chalk
- Coffee filters, any size (3)
- Compass or circular objects of different sizes
- *Computer monitor
- Concentric circles printed on transparency material (see pp. 213–214)
- Cups, small plastic salsa (2)
- Dominoes, coins, or small blocks of wood (similar to Jenga pieces) (8–10)
- Flashlight
- Flashlight bulb, 1.5–3 V
- Food coloring (optional)
- *Glass, drinking, transparent
- Graph paper (2 sheets)
- Graphite artist's pencil, with solid graphite core (may be cut into 4 pieces and sharpened) (Mechanical pencils may be used but do not work as well.)
- *Hole punch
- *Insulators (a variety of materials that the students think up themselves)
- *Lamp
- LED, with current-limiting resistor
- Lightbulb
- *Liquid soap or detergent (a few drops)
- Magnets, small (2) (can be removed from magnetic toys such as magnetic alphabet)
- Magnifying glass
- Marble, large, glass
- Marble, small, glass (5)
- Marble, small, metal
- Marking pen (nonpermanent)
- Mechanical pencil lead refills (0.5 mm and 0.7 mm)
- Multimeter
- Nail, iron
- Oil, cooking

- *Paper (1 sheet)
- Paper clips
- Paper cutout from thick paper (3 cm square)
- Pencil
- Pennies (4)
- Pepper
- Pie pan
- Piezo sparker from inexpensive barbecue lighter
- Ping-Pong ball
- Plastic bag, resealable snack- or sandwich-size
- Plastic box, shoe-box size (the box that holds the materials)
- Plate, paper or Styrofoam, small with rim around the edge
- Protractor
- Push pins
- Quarters (2) (optional)
- Rock
- Rubber band, thick
- Rubber band, thin (2)
- Ruler, plastic with groove down the middle, metric
- *Salt (a pinch)
- Sand, pennies, or other weights
- Sandpaper square
- *Scissors
- Soda bottle, glass or plastic (any size)
- *Soda cans, empty (2)
- Soda can, unopened
- Spoon, large, shiny
- Static electricity materials: PVC pipe (~15 cm long, ½ or ¼ in. diameter), glass (test tube, microscope slide, etc.), balloon, aluminum can, wood, plastic ruler, turkey roasting bag (cut into 15 x 15 cm squares), paper, wool, fur, aluminum foil, silk, salt, pepper, dirt, crisped rice cereal, thread, Christmas tree icicle
- Stopwatch
- Straws, bendable (2)
- String, kite (~30 cm)
- Tape, Scotch or other clear variety
- Thread, small spool
- Thumbtacks (2)
- Tone generator program (free trial download)
- T-pin
- Triple-throw switch (left, middle off, right)
- *Tube television (not a flat screen)
- Vinegar (~125 ml)
- Washers, large (6)
- Water
- Wires (4 pieces, 6 in., stripped both ends; 2 pieces, 12 in., stripped both ends)
- *Wooden plank to use as a ramp

Managing the Boxes

Before assigning take-home labs, consider how you will manage the materials boxes—especially how they will be collected at the end of the school year. Incomplete or lost boxes are a waste of time and money.

Make the distribution of box lids a routine part of the new school year. For example, if the school librarian manages textbook distribution and collection at your school, have that person also assist with the check-in and check-out process for your materials boxes. In my school, when students enrolled in physics and picked up their textbooks from the library, they also signed out the physics lab materials in the form of a box lid. That is, the lids were labeled with bar codes and checked out to students, and the boxes (in my classroom) were labeled with a matching bar code. When a student brought a lid to class, I gave him or her the matching box to be filled with materials. (Tip: Don't put the materials away yet as you will certainly have students checking into your class late. Just put the lids on the boxes of extra materials and store them somewhere convenient.)

Even with such a system in place, you will need to keep careful watch over the boxes. Students may transfer out of the class or the school and take the box with them. If so, send a letter to their new classes (if they are still enrolled in the school), home, or new school asking for the boxes to be returned. If the boxes are still not returned, treat them just like a textbook and charge a fee to cover the loss. Students should not receive transfer grades until all debts are cleared, including the boxes. It is always preferable—and less costly—to get the box back than to receive money for a new box.

Tell students at the beginning of the year that the boxes must be returned at the end of the year and that some items will be used up while others will still remain. Also tell them approximately how much each item is worth in case they are lost or broken. The librarian, or whoever collects textbooks at the end of the year, can charge these partial damages just like damage to a book. After students return the boxes, the teacher determines if all the materials are there and takes the lids and a list of missing items back to the library.

Do not wait until the last day of the school year to collect the boxes, as students inevitably will be absent or forget to bring their boxes. Rather, start collecting the boxes as soon as the last lab is finished. This will give you several weeks to collect them.

Safety

Every lab in this book is safe when the directions are followed properly. No dangerous chemicals, flammables, fire, or explosives are used. But there is always a chance of accidental injury. Teachers should always remind students to perform activities safely. If your policy is to require goggles with every lab, a pair of goggles can be provided for a small additional cost. Suitable impact-resistant goggles may be purchased at hardware stores.

There are possible choking hazards for very small children. Emphasize to your students that the boxes of lab materials *must* be kept out of the reach of small children. On the next page is a label that should be on the top of all boxes before they are distributed. Just print the page onto peel-and-stick paper, cut out the stickers, and place a sticker on each box or lid. If the peel-and-stick paper is made for the computer printer, you may need to scan the page first.

See page xxii for a letter to parents that explains the possible dangers of the box of lab materials. This letter should be sent home, signed, and returned.

WARNING!

KEEP OUT OF REACH OF SMALL CHILDREN. KEEP THE LID ATTACHED FIRMLY AT ALL TIMES AND THE BOX ON A HIGH SHELF. CHOKING AND SWALLOWING HAZARDS FOR SMALL CHILDREN CONTAINED INSIDE.

WARNING!

KEEP OUT OF REACH OF SMALL CHILDREN. KEEP THE LID ATTACHED FIRMLY AT ALL TIMES AND THE BOX ON A HIGH SHELF. CHOKING AND SWALLOWING HAZARDS FOR SMALL CHILDREN CONTAINED INSIDE.

WARNING!

KEEP OUT OF REACH OF SMALL CHILDREN. KEEP THE LID ATTACHED FIRMLY AT ALL TIMES AND THE BOX ON A HIGH SHELF. CHOKING AND SWALLOWING HAZARDS FOR SMALL CHILDREN CONTAINED INSIDE.

WARNING!

KEEP OUT OF REACH OF SMALL CHILDREN. KEEP THE LID ATTACHED FIRMLY AT ALL TIMES AND THE BOX ON A HIGH SHELF. CHOKING AND SWALLOWING HAZARDS FOR SMALL CHILDREN CONTAINED INSIDE.

Dear Parent or Guardian:

Soon your child will be bringing home a box of materials to perform physics labs at home. These activities will give your child more skill in performing labs, analyzing data, and creating charts and graphs, and they will free up more time in class to cover other important topics.

Although all of the labs are safe, there is always a chance of an accident. Please encourage your child to always perform these activities safely and to follow all directions. I assure you that there are no dangerous chemicals, explosives, flammables, sharp objects, or poisons contained in the materials, but there are small objects that could present a choking hazard for small children (such as marbles and a rubber ball).

The box will have a warning sticker on top to discourage young children from opening it. Students should keep the lid on the box at all times and keep the box out of reach of children.

Sincerely,

I have read and understand that if performed properly, these labs are safe. Abuse of these labs could possibly lead to harm. I do not hold the teacher, school, author, or publisher responsible for injuries sustained while performing these labs.

____

Printed Name

____

Signature Date

Acknowledgments

Most of the ideas in this book have some basis in my experience as a teacher and teacher educator. Although ideas cannot be copyrighted, it is my intent to try to provide credit where credit is due for the ideas. The books and websites that are listed below are valuable resources, and the ideas for many of the labs may have originated in them.

Ehrlich, R. 1997. *Why toast lands jelly-side down: Zen and the art of physics demonstrations*. Princeton, NJ: Princeton University Press.

Gibbs, K. 1999. *The resourceful physics teacher: 600 ideas for creative teaching*. Bristol, PA: Institute of Physics Publishing.

Jargodzki, C. 2001. *Mad about physics: Braintwisters, paradoxes, and curiosities*. New York: John Wiley.

Liem, T. 1987. *Invitations to science inquiry*. Lexington, MA: Ginn Press. *www.eric.ed.gov/ERICDocs/data/ericdocs2sql/content_storage_01/0000019b/80/1e/27/27.pdf*.

Minnix, R. B., and D. R. Carpenter. 1993. *The Dick and Rae physics demo notebook*. Lexington, VA: Dick and Rae.

Robinson, P. 2002. *Conceptual physics laboratory manual*. Needham, MA: Prentice Hall.

UNESCO. 1962. *700 science experiments for everyone*. Garden City, NY: Doubleday.

Walker, J. 1975. *The flying circus of physics*. New York: John Wiley.

Woolf, L. 2000. *The line of resistance (Teacher demonstration kit)*. Madison, WI: Institute for Chemical Education.

References

Banchi, H., and R. Bell. 2008. The many levels of inquiry. *Science and Children* 46 (2): 26–29.

Bell, R. L., L. Smetana, and I. Binns. 2005. Simplifying inquiry instruction. *The Science Teacher* 72 (7): 30–33.

Colburn, A. 2000. An inquiry primer. *Science Scope* 23 (6): 42–44.

Dochy, F., M. Segers, and M. M. Buehl. 1999. The relation between assessment practices and outcomes of studies: The case of research on prior knowledge. *Review of Educational Research* 69 (2): 145–186.

Herron, M. D. 1971. The nature of scientific enquiry. *The School Review* 79 (2): 171–212.

Langer, J. A. 1984. Examining background knowledge and text comprehension. *Reading Research Quarterly* 19 (4): 468–481.

Lee, O., S. Fradd, X. Sutman, and M. K. Saxton. 2001. Promoting science literacy with English language learners through instructional materials development: A case study. *Bilingual Research Journal* 25 (4): 479–501.

Llewellyn, D. 2005. *Teaching high school science through inquiry: A case study approach*. Thousand Oaks, CA: Corwin Press.

Marzano, R. 2004. *Building background knowledge for academic achievement: Research on what works in schools*. Alexandria, VA: Association for Supervision and Curriculum Development.

Marzano, R. 2006. *Classroom assessment and grading that work*. Alexandria, VA: Association for Supervision and Curriculum Development.

McComas, W. 2005. Laboratory instruction in the service of science teaching and learning: Reinventing and reinvigorating the laboratory experience. *The Science Teacher* 72 (7): 24–29.

Oleck, J. 2006. Is homework necessary? SLJ talks to Alfie Kohn. *School Library Journal* (December 6). *www.schoollibraryjournal.com/article/CA6397407.html.*

Stevens, K. C. 1980. The effect of background knowledge on the reading comprehension of ninth graders. *Journal of Reading Behavior* 12 (2): 151–154.

SECTION 1:

Motion and Kinematics

LAB 1: DISTANCE VERSUS TIME GRAPHS 1

This inquiry activity should be used before students learn about velocity and distance versus time graphs. Students will discover how the slope of a distance versus time graph is related to the speed of the object.

Topic: Graphing Speed, Velocity, Acceleration
Go to: *www.sciLINKS.org*
Code: THP01

Post-Lab Answers

1. The larger (steeper, higher) the slope, the higher the speed.
2. Answers will vary, but should be less than 100 cm/s. The units are cm/s. The slope represents the speed of the marble.
3. See graph below.

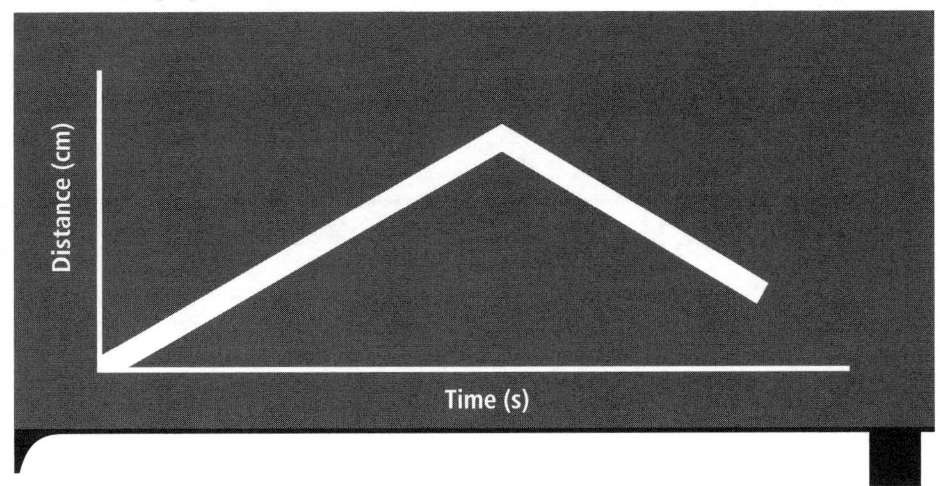

LAB 1: DISTANCE VERSUS TIME GRAPHS 1

QUESTION

How does a distance versus time graph look for an object traveling at constant speed?

SAFETY

Do not leave the marble in a place where someone might step on it or a child might swallow it.

MATERIALS

Plastic metric ruler with groove down the middle, stopwatch, marble (any)

PROCEDURE

In this lab, you will be drawing distance versus time graphs for objects traveling at different constant speeds. The shape and trend of the graph will give you information about other distance versus time graphs you'll see in physics.

1. Put removable marks on a smooth, flat surface (such as a floor or table) at 0, 15, 30, 45, 60, and 75 cm.
2. Using your ruler set up a ramp to roll the marble down. Don't put the bottom of the ramp right on the start line because you want to let the marble stop bouncing before you start timing it.
3. Place the ramp at a low angle (around 30°) and roll the marble from the top of the ramp. Start the watch when the marble crosses the start line and stop it when it gets to 15 cm. Repeat this three times and average the results. You may put an object at the finish line that will make a loud sound when the marble hits it. This will help you stop the watch at the right time.

4. Repeat Step 3, letting the marble roll to 30, 45, 60, and 75 cm. Record all your data in a chart similar to the one below.
5. Repeat the procedure with a higher ramp placed at approximately 45°. Record your data in another data chart.

Distance	Time 1	Time 2	Time 3	Average Time
15 cm				
30 cm				
45 cm				
60 cm				
75 cm				

6. On the same axes, draw two line graphs with average time on the horizontal axis and distance on the vertical axis. Be sure to label the axes of your graph.

Post-Lab Questions

1. What is the relationship between the slope of the graph and the speed of the marble?
2. Calculate the slope of the two lines. What are the units of the slope of the line? What does the slope represent?
3. What would the graph in Step 6 look like if the marble approached a wall at a medium speed, collided with a wall at 45 cm, and bounced off and rolled the opposite direction at low speed? Sketch the line on your graph with a dashed line or in a different color and label it #3.

Extension

Have someone sketch some distance versus time graphs for you and then practice walking in such a way that you would produce that graph. See the example below:

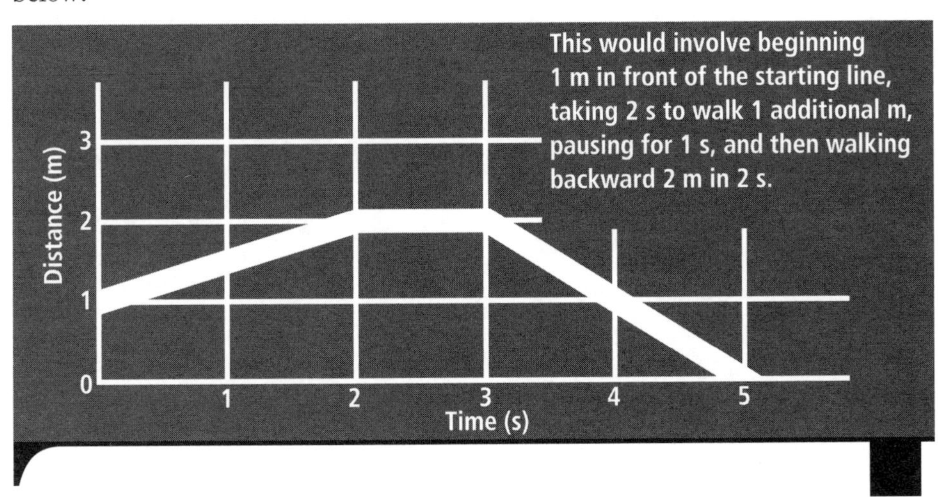

LAB 2: DISTANCE VERSUS TIME GRAPHS 2

This inquiry activity should be performed before students have learned about acceleration but after they have learned about speed. Students should have already completed distance versus time graphs for objects traveling at constant speed (see Lab 1). Some students may know the term *tangent line* and others may not. Teachers are free to change the lab if so desired and should have students create their own data charts. Students should always create their own axes and graphs for experimental data.

Post-Lab Answers

1. The shape is an upward curve. If the object had been slowing down, the shape would be a downward curve.
2. Answers will vary. The first slope will be smaller than the second. This shows that an object travels faster when dropped from a greater height—hence it is accelerating.
3. It takes longer to travel the first half than the last because it is accelerating. Students should see from either their data charts or their graphs that the speed of the marble increases as it falls. A higher average speed over an equal distance will result in a shorter time for the second half.

LAB 2: DISTANCE VERSUS TIME GRAPHS 2

QUESTION

How does a distance versus time graph look for an object that is accelerating?

SAFETY

Do not leave the marble in a place where someone might step on it or a child might swallow it.

MATERIALS

Marble (any), ruler, stopwatch, pie pan or other household item that will make a loud noise when hit by a marble

PROCEDURE

A distance versus time graph for an object traveling at a constant speed is a straight line and the slope of that line is the speed. An object that is accelerating has a changing speed or velocity—hence a changing slope. To figure out the speed of the object during its acceleration, one must take the slope of the tangent line to the graph (see Figure 2.1).

1. Put removable marks on a door frame or outside wall so as not to permanently damage walls in the house. Put the marks at 0.75 m, 1.25 m, 1.75 m, and 2.25 m.
2. Have someone help you drop a marble from each of these heights and into a pie pan or some other object that will make a loud sound when the marble hits. Drop the marble from each of the marks three times and average the results. Come up with a method for coordinating the drop and starting the watch. Use *3–2–1* or *ready–set–go*, but keep it the same each time.

Figure 2.1

Graph of Accelerating Object With Tangent Line

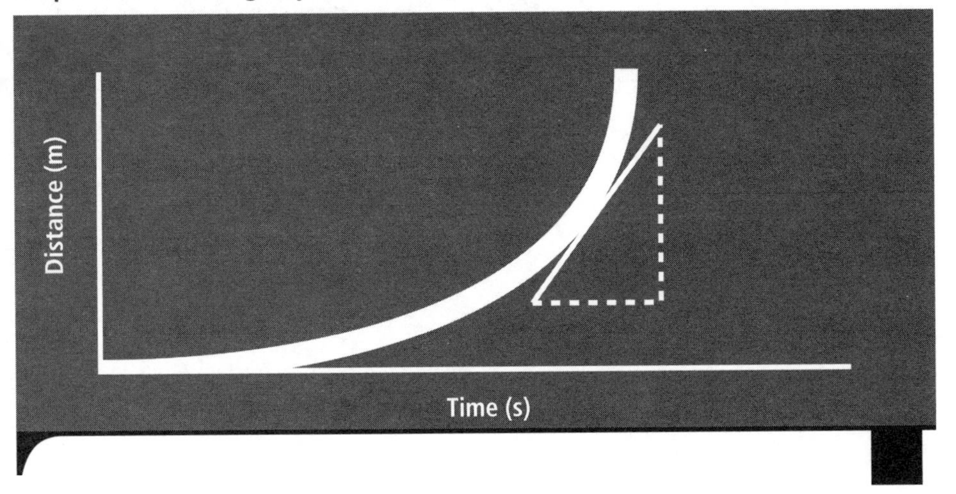

3. Record your data in a chart similar to the one below.

Data Chart

Distance	Time 1	Time 2	Time 3	Average Time
0.75 m				
1.25 m				
1.75 m				
2.25 m				

4. Create a line graph with average time on the horizontal axis and distance on the vertical axis. When connecting the points to make it into a line graph, do it in such a way that it makes a smooth curve. Be sure to label the axes of your graph.

Post-Lab Questions

1. What is the shape of your graph? What would the shape be if the object were slowing down instead of speeding up?
2. Choose a place near the beginning of the graph and draw the tangent line. Choose a point near the end of the graph and do the same. What can you say about the speed of the object as it is dropped from greater heights?
3. When the marble was falling from 2.25 m, did it take longer to fall the first half of the distance or the second half? Explain how you figured this out.

Extension

Go to an amusement park and sketch what a distance versus time graph would look like for different rides based on your estimate of the speed. Alternately, find online videos of roller coasters and do the same. When sketching a graph, there is no need to worry about exact distances; just estimate when the object is going faster and slower and draw the graph accordingly.

LAB 3: AVERAGE SPEED

This lab is not an inquiry activity but will help students understand the meaning of average velocity. Students do not easily understand from a textbook that an object's velocity changes over a period of acceleration. They have difficulty understanding why the final velocity of an object starting from rest is double the average velocity when accelerating at a constant rate. Seeing objects accelerate from rest will help students visualize this concept and will help them solve book problems correctly. Drawing the graphs also will help them visualize this process. The graphs also set up students for an understanding of terminal velocity. It is expected that students are familiar with the definition of *acceleration* and the $a = \Delta V/\Delta t$ equation.

Topic: Speed
Go to: *www.sciLINKS.org*
Code: THP02

Post-Lab Answers

1. The marble accelerated constantly the entire path in Lab 2. The coffee filters stopped accelerating at some point. Or the marble's graph was curved the entire way in Lab 2 whereas the coffee filter's graph flattened out at some point.
2. The graph is not a straight line because the object is accelerating as it falls. The farther it falls, the faster it travels, so the more steeply the line curves.
3. No. Eventually, the object would stop accelerating because the frictional forces will balance out the gravitational forces. The velocity at which this balance occurs is called *terminal velocity*.

LAB 3: AVERAGE SPEED

QUESTION

What can you tell about an object's motion by looking at its distance versus time graph?

SAFETY

Standard safety precautions apply.

MATERIALS

3 coffee filters, ruler, stopwatch

PROCEDURE

In this lab, you will investigate average speed. You will be dropping coffee filters from different heights and measuring the time that it takes for them to hit the ground. *Average speed* is defined as the change of position over a certain segment of time. How quickly something moves a certain distance is the layman's definition. The equation for average speed is:

$$\textit{average speed} = \frac{\textit{change in position}}{\textit{change in time}}$$

If an object travels 5 meters in 2 seconds, it has an average speed of 2.5 m/s. Remember, that it is just an average. This doesn't mean that the object started out or finished at this speed. For example, if you drove 80 mi./hr. for an hour, then rested for an hour, then traveled 80 mi./hr. for another hour, your average speed was 53 mi./hr. even though you may not have traveled 53 mi./hr. for more than a few seconds.

1. Put removable marks on a door frame or outside wall so as not to permanently damage walls in the house. Put the marks at 60 cm, 90 cm, 120 cm, 150 cm, and 180 cm.
2. Have someone help you with dropping or timing the fall of the coffee filters (3 stacked together). Drop the coffee filters from each of the marks three times and average the results. Drop them like an upside down parachute (with the opening facing up). Come up with a method for coordinating the drop and starting the watch. Use *3–2–1* or *ready–set–go*, but keep it the same each time.
3. Record your data in a chart like the one below.

Data Chart

Distance	Time 1	Time 2	Time 3	Average time	Average speed
60 cm					
90 cm					
120 cm					
150 cm					
180 cm					

4. Plot a line graph with average time on the horizontal axis and distance on the vertical axis. Use centimeters as your units for distance and seconds as your units for time.
5. Plot a graph of distance on the horizontal axis and average speed on the vertical axis. Your graph should take an entire piece of paper and the line should occupy most of the graph. Your graphs should not look like Figures 3.1 or 3.2. Graphs may be made on graph paper or regular paper with a ruler.

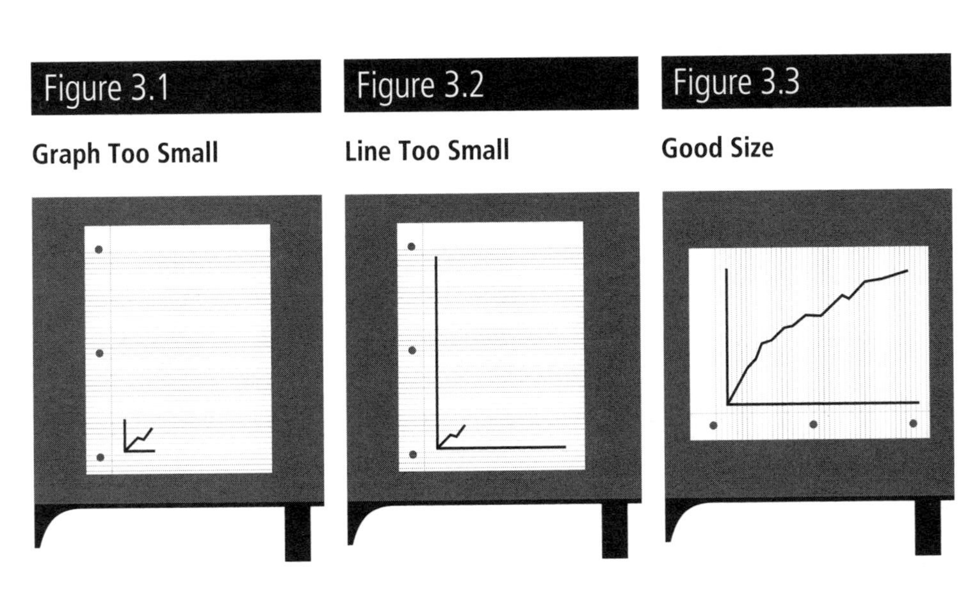

Figure 3.1
Graph Too Small

Figure 3.2
Line Too Small

Figure 3.3
Good Size

Post-Lab Questions

1. What was the difference between the motion of the marble in Lab 2: Distance Versus Time Graphs 2 and the coffee filters in this lab?
2. Why isn't your distance versus time graph a straight line?
3. Do you think that this pattern would go on forever if you kept dropping the filters from higher and higher heights?

LAB 4: FINAL SPEED

This lab is not an inquiry activity. Without a motion sensor, the students cannot determine independently that the final velocity is double the average velocity. This activity does, however, make this concept concrete in their minds. By doing only word problems, students would get the concept but forget it immediately. This activity gives them a hook on which to hang the concept and something concrete with which to connect it.

Post-Lab Answers

1. Both *a* and *c* are starting from zero and accelerating constantly, as required for this relationship. Driving a car involves starts and stops, so this relationship does not apply.
2. Answers will vary. Students should extrapolate their graphs to 3 meters to determine this answer. The theoretical answer should be about 7.7 m/s, but reaction time will have a large effect on answers.
3. The loud sound helps improve reaction time. It is easier to hear exactly when the marble hits than to see it.

LAB 4: FINAL SPEED

QUESTION

How can you use what you know about average speed to determine the final speed of an object starting at rest and accelerating constantly?

SAFETY

Do not leave the marble in a place where someone might step on it or a child might swallow it.

MATERIALS

Marble (any), ruler, stopwatch, pie pan

PROCEDURE

In a previous lab, you learned how to calculate the average speed of an object. The average speed is useful if you're driving and want to know how long your trip will take. Other times it's not useful, like if you want to know your top speed during a drag race or the speed at the bottom of a roller coaster drop. In both of those cases, you want to know your *final speed.*

When you average two numbers together, the average is halfway between them. If the first number is zero, then the second number must be twice as large as the average. For 0 and 10,the average is 5 and 10 is double 5. For 0 and 80, the average is 40 and 80 is double 40.

So, to convert an average speed to a final speed for an object starting from rest and accelerating constantly, all you have to do is double the average speed. This only works when an object is accelerating at a constant rate (like when it's falling) and isn't moving at the beginning of the trial.

Example: A car completes a 400 m drag race in 12 seconds while accelerating constantly the entire race. What was its speed as it crossed the finish line?

$$\textit{average speed} = \frac{\textit{change in position}}{\textit{change in time}} \text{ and } \textit{final speed} = (2)(\textit{average speed})$$

average speed = 400 m/12 s = 33.3 m/s and final speed = (2)(33.3 m/s) = 66.6 m/s

1. Put removable marks on a door frame or outside wall so as not to permanently damage walls in the house. Put the marks at 1 m, 1.5 m, and 2 m.
2. Have someone help you drop a marble from each of these heights into a pie pan or some other object that will make a loud sound when the marble hits but will not interfere with the measurements. Drop the marble from each of the marks three times and average the results. Come up with a method for coordinating the drop and starting the watch. Use *3–2–1* or *ready–set–go*, but keep it the same each time.
3. Record your data in a chart similar to the one below.

Data Chart

Distance	Time 1	Time 2	Time 3	Average Time	Average Speed	Final Speed
1 m						
1.5 m						
2 m						

4. Create a line graph with final speed on the horizontal axis and distance on the vertical axis. Leave room for the vertical axis to go out to 3 m.

Post-Lab Questions

1. For which of the following would the relationship between average speed and final speed that you learned in this lab be valid?
   a. An airplane accelerates constantly on the runway before it takes off.
   b. A car drives from Los Angeles to Las Vegas.
   c. A water balloon falls from a second-story window.
2. According to your graph, what would the speed be if you had dropped the marble from 3 m?
3. Why were you instructed to put a pie plate or other loud object for the marble to fall into? In other words, how does the loud sound help?

LAB 5: ACCELERATION OF GRAVITY 1

This lab is a Level 2 inquiry activity that should be taught after students understand acceleration but before they learn the equations for calculating the acceleration of a body dropped from a certain height ($d = ½at^2$). After this activity, students should understand the significance of the ½ and the t^2 in the equation.

Topic: Force of Gravity
Go to: *www.sciLINKS.org*
Code: THP03

Post-Lab Answers

1. Answers will vary. Answers within 10% should be considered good because of the timing method used. Teachers should remember that the point is not to get the exact answer, but to understand the calculation. Good lab technique and calculations should be rewarded here.
2. The two answers should be very close together. If students understand acceleration well, they should have predicted this. Acceleration of gravity does not depend on height at these small differences.
3. Sources of error include synchronizing the starting and stopping of the stopwatch with the falling of the marble, measuring the height of the marble accurately, and repeating each trial only three times.
4. No, it would not apply here because it was not under constant acceleration, nor did it necessarily begin from rest. It was starting and stopping along the way. It is important to note that this equation is generally true only under conditions of constant acceleration. In the car example, acceleration was not constant.

LAB 5: ACCELERATION OF GRAVITY 1

QUESTION

How can you use the average speed of an object starting from rest and accelerating constantly to determine its acceleration?

SAFETY

Do not leave the marble in a place where someone might step on it or a child might swallow it.

MATERIALS

Marble (any), ruler, stopwatch, pie pan

PROCEDURE

In this lab, you will calculate the acceleration of gravity by timing a marble as it falls. The key to the calculation is realizing that the marble is starting from a speed of zero and accelerating constantly to a certain final speed. The average of these two numbers is what is calculated by taking the distance traveled divided by the fall time. Because you know the average speed and the starting speed, you can calculate the final velocity. You will get more information about this in Lab 4: Final Speed. From the final speed and the time, you can calculate the acceleration.
For an object starting at rest and accelerating constantly:

$$S_{avg} = \frac{(S_i + S_f)}{2}$$

Where S_{avg} = average speed, S_i = initial speed, and S_f = final speed so, if $S_i = 0$, then $S_f = 2(S_{avg})$

1. Put a temporary mark on a wall at a height of 2 m. Put something on the ground that will make a loud noise when the marble hits it. (A pie pan works well for this.)
2. Drop the marble three times and record the times in a data chart like the one below. Calculate the average time.
3. Calculate the final speed of the marble just before it hits the ground and record it in your data chart.
4. Calculate the acceleration of the marble as it falls and record it in your data chart. Recall that under constant acceleration starting from rest, $A = S_f/T$ where A = acceleration and T = change in time or elapsed time.
5. Repeat the experiment again from a different height (do not go lower than 1 m).

Data Chart

Height	Time 1	Time 2	Time 3	Average Time	S_f	Acceleration
2 m						

Post-Lab Questions

1. What was your percent error for the acceleration of gravity? (Take 9.81 m/s^2 as the accepted value.)
2. Were your answers for the two different heights within 20% of each other? At these scales, does the acceleration of gravity seem to depend on height?
3. What were your biggest sources of error in this lab?
4. Would this average speed equation apply to a car making a trip across the country?

Extension

Would you expect to get the same acceleration for any object? Try other objects and see.

LAB 6: ACCELERATION OF GRAVITY 2

This lab is between a Level 2 and Level 3 inquiry activity in that it should be assigned after students understand acceleration but before they learn about the acceleration of gravity. Because there are many sources of error when using a pendulum, students cannot be expected to determine the entire procedure on their own. Tips are given to reduce these errors without dictating the procedure. After this activity, students should clearly understand one way to determine the acceleration of gravity and have more experience using the pendulum equation.

Accurately measuring the length of the pendulum is one of the largest sources of error. Students should understand center of mass before performing this experiment.

Topic: Acceleration
Go to: *www.sciLINKS.org*
Code: THP04

Post-Lab Answers

1. Answers will vary. This method should result in fairly accurate results if errors are reduced. Answers within 10% should be considered good.
2. It is likely that this lab will result in more accurate answers than the previous experiment because of the ease of measuring 10 swings of a pendulum rather than the fall of a marble.
3. The length of the pendulum will not matter. The longer pendulum should be easier because the period is longer. Students sometimes have a difficult time understanding what "significantly different" means. If a pendulum that is 1 m long comes out 9.78 and a pendulum that is half a meter long comes out 9.76, students think they are different. They need practice working with numbers like these and seeing that if a large difference in length made a very small difference in acceleration, it was more likely because of measurement errors.

LAB 6: ACCELERATION OF GRAVITY 2

QUESTION

How are the motion of a pendulum and the acceleration of gravity related?

SAFETY

Standard safety precautions apply.

MATERIALS

Thread, paper clips, washers, stopwatch

PROCEDURE

In this lab, you will be calculating the acceleration of gravity by timing a pendulum. The period of a pendulum is proportional to the square root of the length of the pendulum divided by the acceleration of gravity. Therefore, if you measure the period of the pendulum (the time that it takes for one back-and-forth swing) and measure the length of the pendulum, you can calculate the acceleration of gravity fairly accurately. The equation is shown below (where T is the period measured in seconds, L is the length measured in meters, and g is the acceleration of gravity measured in m/s^2).

$$T = 2\pi \sqrt{\frac{L}{g}} \quad \text{or} \quad g = \frac{4\pi^2 L}{T^2}$$

Determine the period of the pendulum accurately by following these tips:

1. Use paper clips and washers for the pendulum.
2. Choose a length that will give a large period, at least 50 cm.

3. To measure the length of the pendulum, you must measure to the center of mass of the pendulum bob (the washers). Measure the length from where the string is suspended to the center of the washers and record it in a data chart similar to the one below.
4. Do not swing the pendulum from more than 10–15 degrees.
5. Do not try to time one swing; time multiple swings for more accuracy.
6. Repeat your measurements at least three times and average the results. Record all of the results in a data chart.
7. Repeat one more time with a pendulum of a different length to ensure that length does not change the acceleration.

Data Chart

Length	Time 1	Time 2	Time 3	Average Time	Acceleration

Post-Lab Questions

1. What was your percent error compared to the accepted value of 9.81 m/s^2?
2. How did that percentage error compare to Lab 5: Acceleration of Gravity 1?
3. Did it make a significant difference how long the pendulum is? Was one pendulum easier to use than the other?

Extensions

Four students are discussing how a pendulum would act on the moon. One says, “The period of the pendulum would be the same because gravity wouldn’t pull as hard on the downward swing, but it would be canceled out by not pulling as hard on the upward swing either.” The second student says, “The only thing about the pendulum that would change on the moon is the weight, and the weight of an object does not determine how quickly it falls, so it will swing with the same speed.” The third student says, “Remember the lab we did with the pendulum last year? We determined that the only variable affecting the period of the pendulum is the length. As long as you keep the length the same, the period will be the same.” The fourth student says, “According to the equation for the period of a pendulum, the pendulum’s period depends on L and g. Because g is smaller, the period will be larger.”

Who is correct? Explain why. Calculate the period of the two pendulums in this experiment on the Moon, where the acceleration of gravity is 1/6 of the rate of acceleration on Earth.

LAB 7: REACTION TIME

This lab is not an inquiry activity. There are some students whose reaction times will not allow them to catch a 12 in. ruler. They may use a dowel, stick, strip of cardboard, etc. Although the students are led to believe that the point of the lab is to calculate their reaction time, the goal really is to make them familiar with the calculation and the idea of free fall. The reaction time scenario will engage them in the activity.

Topic: Properties of Sound
Go to: *www.sciLINKS.org*
Code: THP05

Topic: Vision
Go to: *www.sciLINKS.org*
Code: THP06

Post-Lab Answers

1. A dollar bill is 0.155 m long. Using $d = ½at^2$, $t = \sqrt{\frac{2d}{a}}$ therefore $t = 0.18$ s. Average human reaction time is around 0.2 s, so it would take a fast person to accomplish this task. If the dollar bill is bent and air significantly slows its motion, then catching it can be easy.
2. Answers will vary. Reaction to sound should be faster than reaction to sight. The person dropping the bill must ensure that he or she says "go" at just the right time, though.
3. Using $d = ½at^2$, the smallest object that could be caught this way would be 0.18 m.

LAB 7: REACTION TIME

QUESTION

Which is faster, your reaction to a sound or your reaction to a visual stimulus?

SAFETY

Standard safety precautions apply.

MATERIALS

Ruler

PROCEDURE

There is a common parlor trick in which a person holds a dollar bill in the space between the slightly spread-out finger and thumb of another person. That person is told that the dollar bill will be dropped without notice and if he can catch the dollar bill, it's his to keep. Most people do not have a reaction time fast enough to catch the bill.

In this lab, you will be recreating this trick but will be using a ruler so it's possible to calculate your reaction time. You should be familiar with the equation $d = 1/2at^2$, where d is the distance that the ruler falls, a is the acceleration of gravity, and t is your reaction time. You will be doing this both with your eyes open and with your eyes closed with an auditory signal.

Get a partner to help you with this experiment; he or she will drop the ruler.

1. Hold your hand out with your thumb and your forefinger separated by about a centimeter. Have your partner put the end of the ruler between your fingers. Instruct him or her to wait a short time and then drop the

ruler without notice. You should catch it as quickly as possible and record how far the ruler fell in a data chart.

2. Repeat Step 1 four more times and average the results. Record your average in the data chart.
3. Repeat Step 1 five more times, but this time you should have your eyes closed and your partner should say "go" just as he or she drops the ruler. Average the results and record the number in the data chart.

Data Chart

Use the skeleton of a data chart below to create your own chart.

Eyes open:

Trial 1 _____ **m** **Trial 2** _____ **m** **Trial 3** _____ **m** **Trial 4** _____ **m** **Trial 5** _____ **m**

Average _____ **m**

Eyes closed:

Trial 1 _____ **m** **Trial 2** _____ **m** **Trial 3** _____ **m** **Trial 4** _____ **m** **Trial 5** _____ **m**

Average _____ **m**

Post-Lab Questions

1. Measure the length of a dollar bill. What would a person's reaction time have to be to catch the dollar bill?
2. Calculate your reaction times from your averages and the equation on the previous page. Don't forget that all of your measurements have to be in meters. Which was faster, eyes open or eyes closed?
3. If your reaction time was 0.19 s, what is the shortest object that you could catch this way?

Extension

Imagine you're driving a car and see an obstacle ahead. You have to react to that situation, move your foot to the brake, and stop the car. Calculate how far you would travel during *your average reaction time* if your car were traveling at 25 mi./hr., 50 mi./hr., 75 mi./hr., and 100 mi./hr. Give your answers in feet (there are 5,280 ft. in a mile).

LAB 8: TERMINAL VELOCITY

This lab is an inquiry activity in that students have not been exposed to the idea of terminal velocity, though they are using skills that they already have to analyze the balloon's motion. The lab is both a review of graphing and translating distance versus time graphs, as well as an inquiry into terminal velocity leading into a discussion about air resistance. A discussion of terminal velocity is included at the end of the lab to give students a name for the phenomenon they have just witnessed. The pennies will help the balloon fall straight, but washers or other suitable objects of the same mass may be used.

SCI LINKS. THE WORLD'S A CLICK AWAY
Topic: Velocity
Go to: *www.sciLINKS.org*
Code: THP07

Post-Lab Answers

1. Answers will vary but change should occur where the graph went from an upward curve to a diagonal line. If students do not notice the straight line, they will not be able to answer question 3 correctly. The teacher will need to follow up on this question in class.
2. Before the change, it was upward curved and accelerating. After the change, it was a straight diagonal line and traveling at constant velocity.
3. Answers will vary but students should calculate the slope of the line.
4. The mass of the balloon would have been virtually the same, but the air resistance would have been less, so it would take more time to reach terminal velocity.

LAB 8: TERMINAL VELOCITY

QUESTION

What is the terminal velocity of a weighted balloon?

SAFETY

Keep balloons and pennies out of the reach of young children; they pose a choking hazard.

MATERIALS

Balloon, tape, 4 pennies, ruler, stopwatch

PROCEDURE

In this lab, you will be using what you know about distance versus time graphs to analyze a graph that you have not seen before. You will drop a light object and measure the time that it takes to hit the ground. The distance versus time graph will be more complex than those that you have done so far.

1. Blow up a large balloon to full capacity. Holding the balloon with the knot on top, tape four pennies evenly along the bottom so that when dropped, the balloon will fall straight down.

Data Chart

Distance	Time 1	Time 2	Time 3	Average Time	Average Speed	Final Speed
50 cm						
75 cm						
100 cm						
125 cm						
150 cm						
175 cm						
200 cm						

2. Put temporary marks or tape on a wall at distances of 50, 75, 100, 125, 150, 175, and 200 cm.
3. Have a partner help you drop the balloon and time how long it takes to hit the ground.
4. Record your data in a chart similar to the one above.
5. Create a line graph with average time on the horizontal axis and final speed on the vertical axis.

Post-Lab Questions

1. From your graph, determine when there was a change in the balloon's motion (for example, from constant speed to acceleration).
2. Describe the shape of your graph before and after the change and explain what the shape means about the balloon's motion.
3. Choose a segment after the change in the balloon's motion and calculate the speed.
4. If you had not blown up the balloon all the way, would it have taken more time or less time to reach that velocity?

Discussion

When an object falls from a large distance, it reaches a point where air resistance prevents it from traveling any faster. This speed is its *terminal velocity*. The terminal velocity of a skydiver is approximately 120 mi./hr. (53 m/s). An object that is less dense than a skydiver can have a much slower terminal velocity. An object that is more streamlined could have a higher terminal velocity. Once the object reaches its terminal velocity, it cruises at that velocity for the rest of its fall.

Recall that a distance versus time graph for an accelerating object curves upward, and for an object traveling at constant speed, the graph produces a straight, diagonal line.

Extension

Analyze a video of a skydiver to determine the person's terminal velocity once the parachute is open. Use landmarks in the video to estimate distances.

LAB 9: EFFICIENCY

Lab 9: Efficiency

This inquiry activity is used after students have learned how to calculate kinetic and potential energy but before they learn about conservation of energy. It is important for students to repeat the measurement three times because it is not easy to measure the exact bounce height. You may choose to give the approximate mass of the ball so that students see that the mass cancels in the calculation.

Topic: Energy Efficiency
Go to: *www.sciLINKS.org*
Code: THP08

Post-Lab Answers

1. Answers will vary. Generally, the efficiencies get worse as the height increases. The quality of the balls has a large effect on efficiency.
2. Answers will vary. Students may either extrapolate the graph or write the equation for the line to determine the answer.
3. The marble's efficiency varies based on the hardness of the surface on which it is bouncing. Marbles bounce very well on hard surfaces and can compete with the efficiency of the rubber ball.

LAB 9: EFFICIENCY

QUESTION

What is the efficiency of a bouncing Super Ball?

SAFETY

Keep the rubber ball out of the reach of young children; it is a choking hazard. Do not leave the ball in a place where someone might step on it.

MATERIALS

Bouncy ball, ruler, marbles (several different types)

PROCEDURE

In a perfect world, when a ball is dropped and bounced, all its potential energy would be converted into kinetic energy and then back to potential energy again. The ball would return to the same height from which it was dropped. In the real world, some of that energy is converted into heat in the ball, heat in the ground, friction with the air, and sound. The ratio of the energy after the collision to the energy going into the collision is called the *efficiency* and is usually expressed as a percentage.

$$efficiency = \frac{energy\ out}{energy\ in} \times 100\%$$

In this case, we will define "energy in" as the initial potential energy of the ball, and "energy out" as the final potential energy of the ball. Your teacher will provide the mass of the rubber ball so you can calculate potential energies.

Data Chart

Drop Height	Energy In	Bounce Height 1	Bounce Height 2	Bounce Height 3	Average Height	Energy Out	Efficiency

1. Put removable marks on a wall or door frame at 0.50 m, 1 m, 1.5 m, and 2 m.
2. Drop the rubber ball from each of those heights three times and measure how high it bounces. Average the three measurements.
3. Record your data in a chart similar to the one above.

Post-Lab Questions

1. Did your efficiencies get better or worse as the drop height increased?
2. Estimate how high the ball would bounce if it were dropped from 3.0 m. Show your work and explain your reasoning.
3. How do you think the efficiency of a bouncing marble would compare to the rubber ball? Try it and see. Try several different types of marbles and see if there is a difference.

Extension

There is a law in physics that says that efficiency can never be greater than 100%. But if you were to throw the ball downward, it would bounce higher than your hand. Does this violate that law? Explain why or why not.

LAB 10: ACCELERATION AND MASS

Although this lab is not an inquiry activity, it is very important in learning about acceleration and mass. It is a deeply held misconception among students that objects of different masses fall at different rates. Simply explaining that this is not true will not be enough to change that misconception. Teachers must perform demonstrations and students must experience this phenomenon themselves. This activity will help students release the misconception and see how air resistance affects the acceleration of an object.

Teachers should resist calling air resistance "air friction," as it is not friction as students have come to know friction and will lead to misconceptions. *Kinetic friction* (the type of friction that students will imagine you are referring to) is not affected by the speed of an object and air resistance does vary with speed. There is a little friction involved, but most of the slowing of an object passing through air is caused by collisions with air, the compression of air, and the force required to make the air swirl and move.

Teachers can demonstrate to students that the air that a car passes through does not touch the horizontal surfaces of the car by wetting their car and then having a passenger videotape the hood of the car while it's driven on the highway. The droplets of water will creep slowly up the hood of the car even at highway speeds. If air were contacting the car's surface to cause friction, the water would be removed immediately. There is a little bit of friction between this boundary layer of air that surrounds the car and the still air that the car is passing through, but it is small compared to the force of the car's vertical surfaces colliding with the air.

Post-Lab Answers

1. Yes, within a reasonable amount, they will fall at the same speed. Just as in Lab 6: Acceleration of Gravity 2, students need to understand that these marbles are significantly different in mass, so a small difference in speed is due to measurement errors, not physics.
2. The coffee filters will not fall at the same speed because their motion is dominated by friction due to their small mass and large surface area.

3. The patterns in the previous questions are explained by the part of the rule that states, "if friction is not a large factor." With the marbles, friction is not a large factor. With the coffee filters, friction is a large factor.
4. Gravitational force is directly proportional to mass. An object that is twice as heavy will be attracted toward Earth with twice the force. The double mass will be exactly cancelled out by the doubled force because of gravity. In the $F = ma$ equation, if mass is doubled, force will be doubled and acceleration will remain constant.

LAB 10: ACCELERATION AND MASS

QUESTION

Does the acceleration of an object in free fall depend on its mass?

SAFETY

Do not leave the marbles in a place where someone might step on them or a child might swallow them.

MATERIALS

Ruler, small glass marble, large glass marble, stopwatch, 3 coffee filters, pie pan

PROCEDURE

There is a rule in physics that is counterintuitive and many students don't believe it even after reading it in their textbooks. The rule says that all objects, regardless of their mass, fall at the same rate if air resistance is not a large factor. This would mean that a bowling ball and a golf ball would hit the ground at the same time if dropped simultaneously. In this lab, you will be testing that rule to see if it is valid or not.

1. Put a removable mark (such as a piece of tape) on a wall at a height of 2 m.
2. Have someone help you drop objects and time how long it takes them to fall.
3. Drop a small marble from 2 m and time how long it takes to fall three times and average the times. Drop a large marble from the same height and repeat. Adding a pie pan or other metal object to make a sound might help the accuracy of the measurement.
4. Repeat this procedure with coffee filters. First drop one filter, then a stack of two, and then a stack of three.
5. Record your data in a chart similar to the one on page 42.

Data Chart

Object	Time 1	Time 2	Time 3	Average Time	Average Speed
Small marble					
Large marble					
One filter					
Two filters					
Three filters					

Post-Lab Questions

1. Within reason, did the small and large marbles fall at the same speed?
2. Within reason, did the coffee filters all fall at the same average speed?
3. Which part of the rule of falling objects explains each of your previous answers?
4. Newton's famous equation $F = ma$ seems to show that if an equal force is applied to two objects, their accelerations will depend on their masses. How can you use what you know about gravity and this experiment to explain why two objects of different mass can accelerate at the same rate when dropped?

Extension

Drop a book and a sheet of paper side by side. Now place the sheet of paper flat on top of the book and drop them together. Can you explain what happens and why it happens?

LAB 11: INERTIA

This lab is an inquiry activity to be used before discussing inertia. Out of necessity to explain the observation, a short discussion of inertia is included, but teachers should certainly spend more time making this concept concrete. This activity will be the background knowledge on which the rest of the discussion of inertia can be built.

Topic: Inertia
Go to: *www.sciLINKS.org*
Code: THP09

Post-Lab Answers

1. Adding the washers to the center piece of paper gave it more mass and therefore more inertia. The more inertia an object has, the more it tends to stay at rest or resist motion. Its resistance to motion gave the student something to pull against to tear the papers apart.
2. The bug does not have very much mass and therefore not much inertia. It is easy for the car to change the direction of the bug. If it were a flying elephant, however, the passengers would feel that.
3. It is important to wear a seat belt because a person has quite a bit of inertia. A massive object traveling with high speed has a lot of inertia and will continue traveling at high speed when the car stops. Without a seat belt, the person will keep on moving while the car stops upon hitting an obstacle. When the person's forward motion is suddenly stopped after hitting part of the car, he or she will be badly injured. A seat belt provides the force to bring the person's motion to a gradual stop. This is preferable to the sudden stop that results from hitting the steering wheel or dashboard.

LAB 11: INERTIA

QUESTION

How can inertia help you tear a piece of paper in three pieces in midair?

SAFETY

Standard safety precautions apply.

MATERIALS

Sheet of paper, scissors, 3 washers, tape (Scotch or other)

PROCEDURE

You may have seen a challenge in which a person is given a small piece of paper torn in two places and is challenged to tear it into three pieces with one pull. It is very difficult to do until you know more about a physics concept called *inertia*.

1. Cut out a rectangle of paper and put two cuts in it as shown in Figure 11.1. Cut out several more rectangles so that you can try different strategies.
2. Try to tear the rectangle into three pieces with one pull. Try pulling slowly, quickly, up, down, and any way that you think might help.
3. Tape the washers to the center piece and see if you can pull the paper in three by pulling it very quickly. If you do not get it the first time, add more washers and try again.

Figure 11.1

Paper Challenge

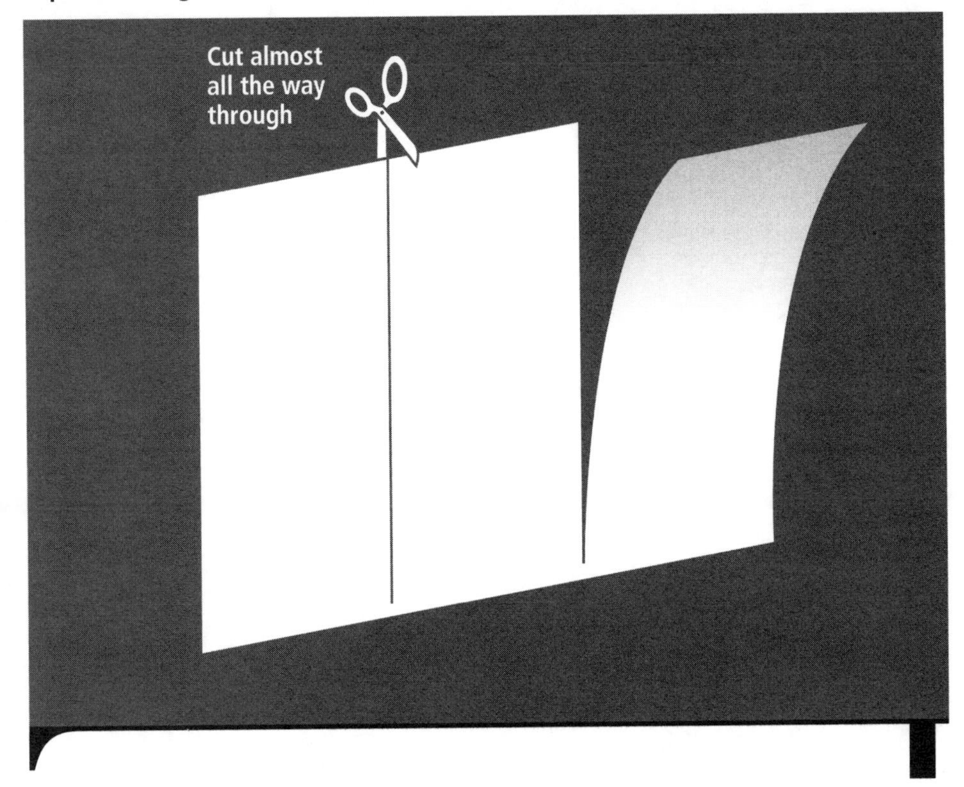

Post-Lab Questions

1. Explain why adding the washers allowed you to pull the paper apart. In your explanation, use the words *mass*, *inertia*, and *rest*. Read the discussion below if you are unfamiliar with any of these terms.
2. Why is it that when a bug hits a car windshield, passengers cannot feel the impact?
3. Using the principles of inertia, explain why it is important to wear a seat belt when traveling in an automobile.

Discussion

To explain how this activity works, you must know a little more about inertia. Inertia is the property of matter that makes objects at rest difficult to move and objects in motion difficult to stop or change directions. The heavier an object is, the more inertia it has. Sometimes you will hear the term *inertia* used in everyday life to explain why a car slid off the road in a sharp turn.

LAB 12: CONSERVATION OF MOMENTUM 1

This lab introduces students to the idea of recoil and how conservation of momentum can be used to explain it. Many people use action/reaction to explain recoil, but conservation of momentum can be used equally well. Before shooting a gun, there is zero momentum relative to the shooter. So after the shot, there must be zero momentum. Whatever momentum the bullet has away from the shooter must be equal and opposite to the momentum of the gun's recoil. This lab is a simple method for demonstrating recoil.

Topic: Law of Conservation of Energy
Go to: *www.sciLINKS.org*
Code: THP10

Post-Lab Answers

1. When I blow through the straw, the air goes to the left and the straw moves to the right.
2. The marble collision is the same thing that happened in the curve of the straw. In order for the straw to redirect the air, it moves the opposite way. If one of the marbles is to be redirected to the left, the other moves to the right.
3. When air is traveling straight down the straw, it has no left-and-right momentum. To turn the 90° corner, it must be given some left-and-right momentum by the straw. When that happens, the straw gets an equal amount of momentum the other way so that the sum of the left-and-right momentum is zero.

LAB 12: CONSERVATION OF MOMENTUM 1

QUESTION

How does conservation of momentum help explain recoil?

SAFETY

Do not leave the marbles in a place where someone might step on them or a child might swallow them.

MATERIALS

Straw, 2 small glass marbles, resealable snack- or sandwich-size plastic bag

PROCEDURE

Both velocity and momentum are vectors. This means that if either velocity or momentum changes direction, then they are not constant even if speed stays the same. It takes a force to change either velocity or momentum of an object. If a soccer ball is rolling directly north and you kick it to the west, you will change its velocity and its momentum.

Because momentum is a vector, the direction of the motion of the object is important. If an object is traveling directly north, it has no momentum in the east/west direction. To get it moving in the east/west direction, another object will have to collide with it and transfer some of its momentum. This is the gist of the law of conservation of momentum. When two objects collide in a closed system, momentum can be transferred from one to the other, but the total amount of momentum remains the same.

1. Bend a clean, flexible straw so that it makes a 90° angle.

2. Put the long end of the straw in your mouth with the opening in the short end pointing to your left.
3. Blow into the straw. Describe what happens in Post-Lab question 1.
4. Put a marble on the floor. Drop another marble so that it hits one side of the marble on the floor. Which way does the dropped marble go? Which way does the marble on the floor go?

Post-Lab Questions

1. What happened in Step 3 when you blew into the straw?
2. How was the marble collision in Step 4 similar to what happened with the straw?
3. Using the idea of conservation of momentum, explain your answer to question 1. You might use the following terms or phrases in your answer: *momentum, change of direction, conservation, opposite direction, add up to zero.*

Extension

Put a resealable bag over the short end of a flexible straw and seal it as well as you can. Blow into the straw. What happens? Can you explain this using conservation of momentum?

LAB 13: CONSERVATION OF MOMENTUM 2

This inquiry activity should be performed after students have learned about momentum, but before they learn about conservation of momentum. Students will discover that when two objects push off each other, the momentum must be equal but in opposite directions. The lab will also reinforce that momentum depends on both math and velocity, which is an idea with which students may struggle.

Post-Lab Answers

1. Answers will vary. If students repeat the experiment several times to determine the correct position, they can expect to get answers within 10% of the actual ratio.
2. The person will not travel backward at 200 m/s because the person is so much more massive than the bullet. If the person is 2,000 times heavier:

   (2,000x)(V) = (x)(200 m/s), where x is the mass of the bullet
   V = 0.1 m/s

3. 100 kg(5 m/s) = 50 kg(V)
   V = 10 m/s

LAB 13: CONSERVATION OF MOMENTUM 2

QUESTION

How does conservation of momentum affect the velocity of dissimilar objects that push off each other?

SAFETY

Do not leave the marbles in a place where someone might step on them or a child might swallow them.

MATERIALS

2 glass marbles, metal marble, plastic ruler with groove, 3 × 5 in. card

PROCEDURE

Momentum is the product of mass and velocity (momentum = [mass][velocity]). In cases where two objects are allowed to move for the same amount of time, distance traveled is directly proportional to speed. A faster object will travel farther than the slower object in equal time. If you compared the momentum of two objects that had traveled for 2 seconds, the 2 seconds would cancel out of both sides of the equation and the only thing that would matter is distance traveled.

In this lab, you will be rolling two objects by having them push off each other (so that momentum must be conserved) and letting them travel for equal times. If they are allowed to move for the same amount of time, then distance may be substituted for velocity in the equation. Because the marbles begin with zero total momentum, they have to end with zero total momentum. By substituting distance for velocity and setting the sum of their momentum to 0, you will calculate the ratio of their masses.

1. Find a flat surface and verify that it is flat by putting the ruler on the surface and placing a marble in the groove. The marble should not move. Tables are not generally level. Floors and countertops work better. You may be able to rotate the ruler around in a circle to find an orientation that is level.
2. Get a 3 × 5 card or playing card and gently fold it in half so that it forms a "V" shape. Notice that when you squeeze the card closed, it springs back open. This is what you will use to launch the marbles (see Figure 13.1).

Figure 13.1

Launching the Marbles

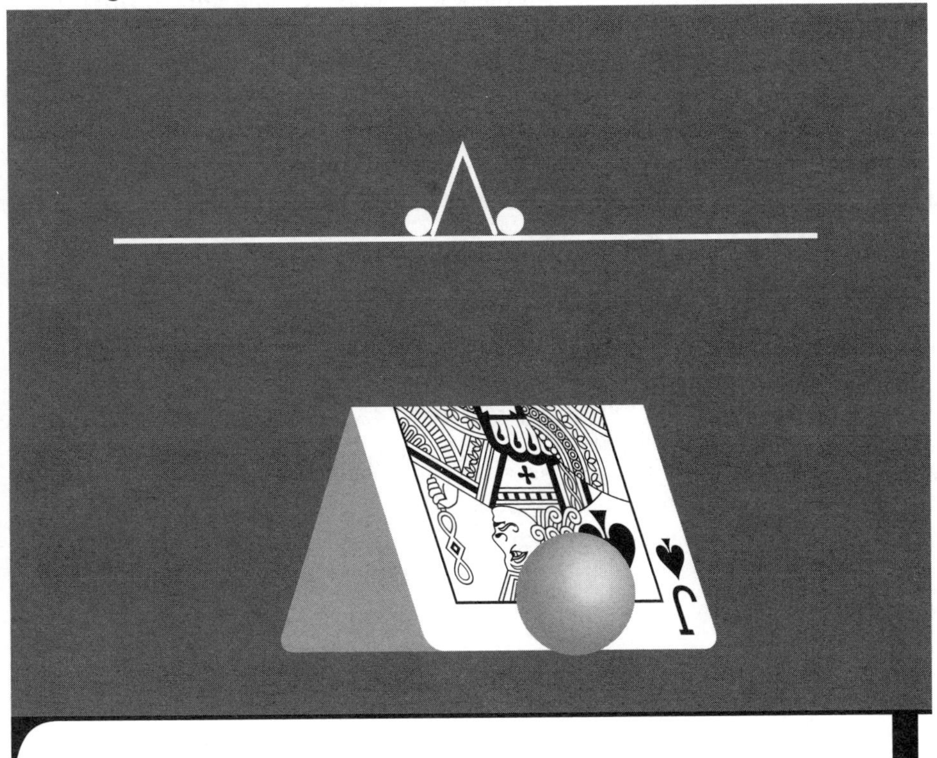

3. To verify that you are operating the launcher correctly, use two of the same marbles and launch them from the center of the ruler. The marbles should reach the ends of the ruler at the same time. This is required to conserve momentum. To add up to 0, they must have equal momentum in opposite directions. If their masses are equal, their velocities must also be equal but in opposite directions.
4. Practice the following procedure several times: Starting with the card in the middle, squeeze the card, put one marble on each side, and release the card. The marbles should take off in opposite directions. Keep in mind these tips as you are practicing:

a. Don't hold the card in your hand, release it. The marbles must push off each other, not your hand.
b. The strength of the launcher doesn't really matter as long as both marbles reach the ends of the ruler.
c. Try to make the marbles touch the card before releasing it. Don't let one marble get a head start.

5. Put the card in the middle of the ruler and release it. Which marble reaches the end of the ruler first?
6. Put the card at the 7.5 cm mark (approximately the one-quarter mark) with the glass marble on the short side of the ruler and the metal marble on the long side of the ruler. Which marble reaches the end of the ruler first?
7. Put the card at the 22.5 cm mark (approximately the three-quarter mark) with the glass marble on the long side of the ruler and the metal marble on the short side of the ruler. Which marble reaches the end of the ruler first?
8. Now move the card back and forth until you find the point at which the two marbles hit the ends of the ruler at the same time. What is the ratio of their distances from their respective ends of the ruler? What is the ratio of the masses of the marbles?

Post-Lab Questions

1. Your teacher will give you the real masses of the marbles. What was the percentage error in your mass ratio?
2. Explain why a person standing on frictionless ice who shoots a bullet at 200 m/s does not fly backward at 200 m/s. Is velocity conserved? Is momentum conserved?
3. Two people are standing near each other on ice. One has a mass of 100 kg and the other has a mass of 50 kg. They push off each other. If the 100 kg person moves backward with a speed of 5 m/s, what is the speed of the 50 kg person?

Extension

Three astronauts of equal mass are floating relatively motionless in space and decide to play a game. Two of the astronauts will push the third one back and forth in a game of catch. How many times can they push the third astronaut before the game ends?

LAB 14: NEWTON'S CRADLE

Topic: Newton's Laws of Motion
Go to: *www.sciLINKS.org*
Code: THP11

This inquiry activity is performed by students after they learn about momentum and energy, but before they discuss conservation of momentum or energy. If students have never seen a Newton's Cradle before, put one on your desk a week before this activity and invite them to "play" with it. If you don't have one, there are applets online that simulate the activity of the device.

Many descriptions of how this device works say that if you swing one ball, then the only way for both momentum and kinetic energy to be conserved is if one ball comes out the other side. This is not true. For example, take the case of a three-ball system with balls of equal mass. If one ball is launched with a velocity of 6 units and a mass of 1 unit, then it has a momentum of 6 and a kinetic energy of 18. What is observed is that one ball comes out with a velocity of 6 units, giving it a momentum of 6 and a kinetic energy of 18. But is that the only solution? No. If the original ball were to rebound with a velocity of -2 units and each of the other two balls traveled away at 4 units, both momentum and energy would be conserved. The total momentum would be $4 + 4 - 2 = 6$ and the kinetic energy would be $2 + 8 + 8 = 18$. Another example would be if the original ball rebounded with a velocity of -0.85 units and one ball came out at 1.00 unit and the other at 5.85 units, both momentum and kinetic energy would be conserved.

There are an infinite number of possible outcomes for this device, not just one outcome as most explanations contend. See *www.lhup.edu/~dsimanek/scenario/cradle.htm* for a complete description of how this demonstration works.

In the procedure, the marbles act exactly like the commercially available Newton's Cradle. If one marble comes in, one goes out. If two come in, two go out. If three come in, three go out, and so on.

Post-Lab Answers

1. If the original ball stops, then the outgoing velocity will have to be v to conserve momentum.
2. Assuming that the marbles have equal mass and that students judged that the velocities were equal before and after, they can show that $mv = mv$ and that $2mv = 2mv$ and that $3mv = 3mv$ so momentum was conserved in each case.

Section 1

3. Assuming that the original ball comes to rest, if one ball came in with a velocity of v and two balls came out with velocities of $1/2v$, the incoming kinetic energy would be $\frac{1}{2}mv^2$ and the outgoing would be $\frac{1}{2}m(0.5v)^2 + \frac{1}{2}m(0.5v)^2$ and these are not equal. If we assume a mass of 1 and a velocity of 10, before = 50 and after = 25.

LAB 14: NEWTON'S CRADLE

QUESTION

How is conservation of momentum involved in a Newton's Cradle?

SAFETY

Do not leave the marbles in a place where someone might step on them or a child might swallow them.

MATERIALS

Ruler, 5 marbles (any kind)

PROCEDURE

A Newton's Cradle is a common desk toy that has five or six balls hanging from fine strings. The balls can be set into motion and their behavior observed. Most explanations that you'll read on the internet or other sources have flawed arguments that say that the action that you see is the only possible action, but that is simply not true. To make the explanation more accurate, these statements should be qualified with, "If it is assumed that the original moving ball comes to rest and does not bounce, then the outcome is the only outcome possible." This is the observed behavior and is a more accurate statement.

You will be building a version of a Newton's Cradle that does not use any strings. You will use it to analyze the conservation of momentum and conservation of energy.

1. Lay five marbles side by side and touching each other in the groove of the plastic ruler.

2. From one end, shoot a marble so that it hits the other four marbles. Observe what happens.
3. Now repeat Step 2, but this time push two marbles. Observe what happens.
4. Now repeat Step 2, but this time push three marbles. Observe what happens.
5. Now repeat Step 2, but this time push four marbles. Observe what happens.
6. Now roll one marble extra fast. Can you get two marbles to shoot out the other side at half the speed?
7. Try other combinations to try to get a different number of marbles to come out than the number that went in.

Post-Lab Questions

1. If a ball of mass m and velocity v comes in and stops and an identical ball flies out the other side, what will its velocity have to be to conserve momentum?
2. Show that momentum was conserved in each of the collisions in Steps 2–7.
3. In Step 6, show that although momentum would be conserved if two marbles rolled out half as fast (to conserve momentum), kinetic energy would not be conserved because the original ball came to rest ($KE = \frac{1}{2}mv^2$).

Extension

You can perform a similar experiment with coins. Put a row of coins on a smooth surface and launch a coin at the row. Try putting your finger on one of the coins to keep it from moving and see if it still works. Try stacking coins on top of each other in the middle of the row and see if anything changes. Again try to do anything you can to make a different number of coins shoot out of one side than come in from the other side.

LAB 15: INDEPENDENCE OF VELOCITY

This inquiry activity should be completed before discussing with students that a projectile's motion in the vertical direction is independent of its motion in the horizontal direction. As long as students use their apparatus carefully and don't flip coins in the air or launch them at an angle, they will clearly see that a falling coin and a projected coin will hit the ground at the same time. There are many apparatus available for demonstrating this, but allowing students to do it on their own will result in better understanding and longer retention.

Post-Lab Answers

1. They will hit at the same time. A small number of students may not get this result because of problems constructing or activating their devices, so it is important that the teacher clarify the intended result and use other devices to demonstrate the same effect after the students have completed this lab. (Try Ping-Pong ball guns, toy crossbows, and apparatus available from science catalogs.)
2. Height doesn't matter; the coins always hit the ground at the same time.
3. Gravity accelerates all objects at 9.8 m/s^2 vertically, independent of whether the object is moving horizontally.
4. If the corners were higher, the launched coin would have been shot upward, resulting in a higher hang time, and then woul hit the ground after the dropped coin. If the corners were lower, it would have had an initial downward velocity and would have hit the ground before the dropped coin.

LAB 15: INDEPENDENCE OF VELOCITY

QUESTION

What factor determines how long a projectile will hang in the air?

SAFETY

Keep push pins out of the reach of young children.

MATERIALS

Ruler, 2 quarters (or washers), push pin, cardboard

PROCEDURE

There is a common question in physics classes about a person who aims a gun horizontally and fires it just as he drops a bullet from the same height. The question asks which bullet hits the ground first (ignoring air resistance). This is a counterintuitive question that many physics students have a difficult time understanding the first time they hear it. You will be simulating this problem by projecting one coin and dropping another at the same time and using the sound that they make to hear which one hits first or if they hit at the same time.

1. Set up the launching apparatus on a flat piece of cardboard (see Figure 15.1). Put your ruler on the cardboard and make a pivot in one end with a nail or thumb tack. Use a push pin to make a pivot point for the ruler. Plastic rulers usually have a hole that you can use. You can push the pin through a wooden ruler.
2. Put one quarter or washer on the ruler such that it hangs over the edge of the cardboard. Another quarter should be placed next to the ruler near the edge of the cardboard.

Figure 15.1

View of the Launching Apparatus From Above

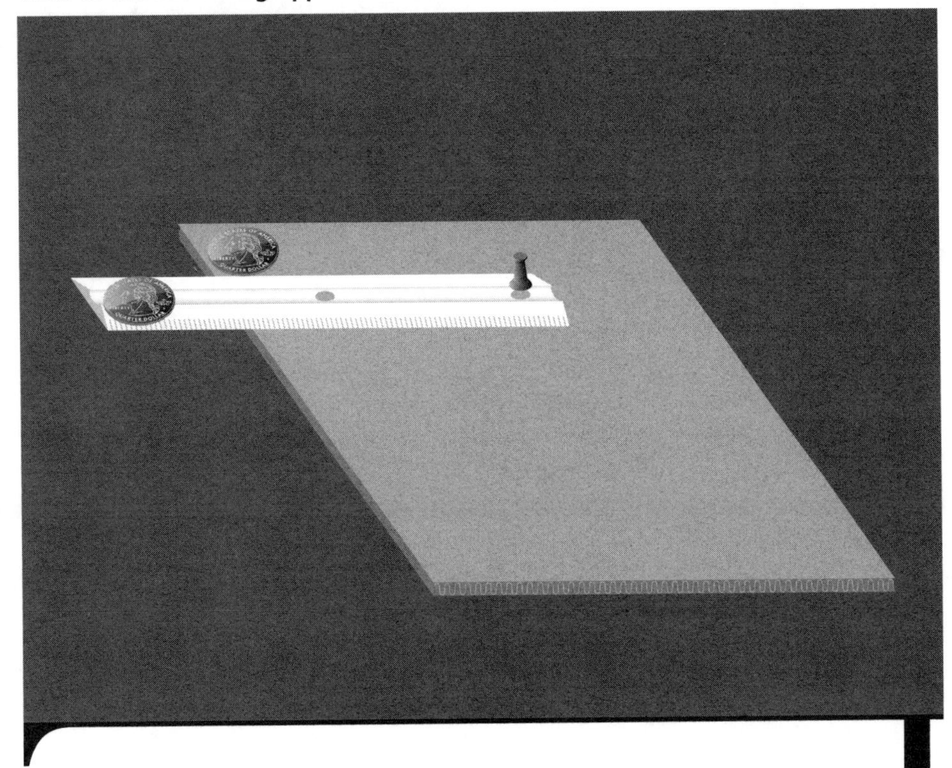

3. Be sure that the entire apparatus is flat on a table, and then with a sharp blow, set the ruler spinning. One quarter will fall straight down and the other will be launched to the side. Try flicking the end of the ruler with a pencil. Record how the quarters hit the ground, either one hitting first or both hitting together.
4. Now raise the apparatus higher and do it again. Raise it higher once more and repeat it again.

Post-Lab Questions

1. Which quarter hit the ground first, or did they hit at the same time?
2. Did the height of the apparatus make any difference?
3. Explain your results in terms of the acceleration of gravity.
4. If the table had been slanted so that the corners were higher than the rest, which coin would have hit the ground first? What would happen if the corners were lower?

Extension

Try this experiment with a toy dart gun or ball gun. Hold the gun as high and as flat as you can, and just as you shoot, drop another ball from the same height.

LAB 16: TORQUE AND FIRST-CLASS LEVERS

This inquiry activity should be done before students have studied the operation of a first-class lever. The activity could be used when discussing simple machines or torque. Although it involves the use of a lever, the activity is not really about simple machines; it is about torque. If students have been taught about levers previously, you may see the "Aha!" moment when they suddenly realize why the sum of the products of mass and distance are equal on each side of the lever. The concept of torque helps them understand why the lever works the way that it does.

Topic: Torque
Go to: *www.sciLINKS.org*
Code: THP12

Post-Lab Answers

1. On the side with more weight, the weights were closer to the fulcrum.
2. The product of the washers and the distance from the fulcrum are equal on both sides. So torque is force times distance.
3. $1 \times 15 = 15 = 1 \times 15$
   $2 \times 10 = 20 = 2 \times 10$
   $2 \times 5 = 10 = 1 \times 10$
   $1 \times 10 = 10 = 2 \times 5$
   $3 \times 10 = 30 = 2 \times 15$

LAB 16: TORQUE AND FIRST-CLASS LEVERS

QUESTION

How can torque be used to explain the workings of a first-class lever?

Standard safety precautions apply.

MATERIALS

Ruler, washers, thread (~20 cm)

PROCEDURE

A first-class lever has the fulcrum in the middle, the object to be lifted on one side, and the force on the other. You will be investigating how different masses can be arranged to balance out a lever. When a force is pushing on a rotating object in a direction that does not pass through its center, this is called a *torque*. When you push on a bicycle pedal, you are applying a torque. When you use a wrench to loosen a bolt, you are applying a torque.

1. Tape a piece of thread near the center of your ruler such that it is balanced when hanging from the thread. You can use small pieces of tape to fine-tune the balance of the ruler because the ruler may have a hole at one end.
2. Open four paper clips that you will use to hang weights (washers) from the ruler. One paper clip should be attached to each end of a short string. One end will hang from the ruler. The other end will hold the washers. The whole string assembly can slide along the ruler to balance the apparatus.
3. Hang one washer from each of two paper clips and put one 15 cm to the right of the center. At what point must you hang the other to the left of the center to get it to balance out?

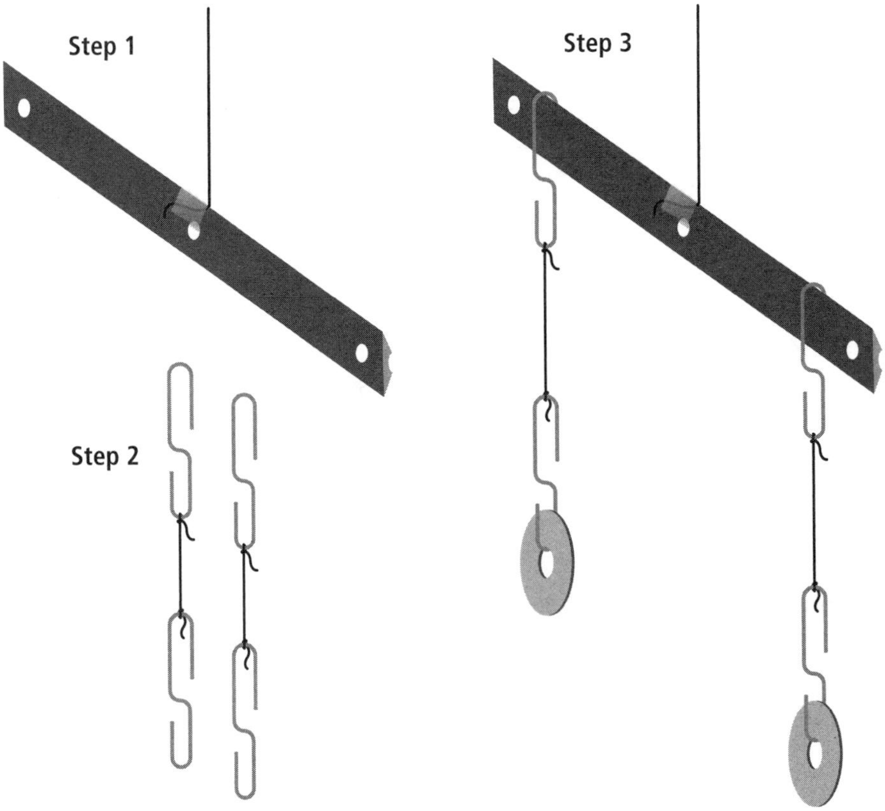

Step 1

Step 3

Step 2

4. Repeat the process with the combinations of washers and distances in the data chart below.

Data Chart

Left Side of Ruler		Right Side of Ruler	
Number of Washers	Distance From Center (cm)	Number of Washers	Distance From Center (cm)
1	15	1	
2	10	2	
2	5	1	
1	10	2	
3	10	2	

Post-Lab Questions

1. On the side of the ruler with more weight, were the weights closer to the center or farther from the center?

2. Torque is related to force and the distance from the center. When a lever is balanced, the torques on either side of the center are equal. Is torque force times distance, distance divided by force, or force divided by distance?
3. Using your equation from #2, calculate the torque on either side of the ruler and put it in the center column of your data chart. Use washer × cm for your units. (Note: Normally, Newton · meters are used to measure torque. Sometimes you will see the nonmetric unit of foot · pounds as well. Because you do not know the mass of the washers, you will use "washers" as the unit for weight.)

Extension

Torque is used frequently with automobiles. It is used to describe the towing power of a truck and also how tight to tighten nuts and bolts when repairing a car. Find out on the internet how many foot · pounds of torque a 2009 Ford F-150 pick-up truck has and convert it to Newton · meters.

LAB 17: WHAT IS A RADIAN?

This is an inquiry activity in which students discover what a radian is. Some students may already know how many degrees are in a radian or how many radians are in a circle, but they rarely know the definition of a radian. Knowing the definition will allow them to better understand angular velocity and angular acceleration.

Post-Lab Answers

1. There are just over 6 radii around the circumference of the circle. That is closest to 2π radians. (We find the exact relationship from the circumference = 2 pi radius relationship.)
2. They were all the same no matter what diameter the circle had.
3. There are 57° in a radian.
4. A radian is the angle subtended by one radius around the circumference of a circle. In student language, a radian is how many degrees it takes for a radius of a circle to go around the outside of the circle.
5. Because we now know that one radius around the circle subtends approximately 57°, just find out how many times larger than 57° the object traveled and multiply that by the radius of the circle.

LAB 17: WHAT IS A RADIAN?

QUESTION

What is a radian?

SAFETY

Keep scissors out of the reach of young children.

MATERIALS

Thread, ruler, protractor, scissors, compass or circular objects of different sizes

PROCEDURE

Most people know how to convert from degrees to radians and back. But do you know what a radian is? A radian is not simply a certain number of degrees. In this lab, you will learn the definition of a radian.

1. Draw a circle on a blank sheet of paper with a compass. If you don't have a compass, trace a circular object on the paper. Put a mark in the center.
2. Cut eight pieces of string the length of one *radius* of this circle. Record the radius in "Radius (cm)" on a data chart like the one on page 69.
3. Count how many of these stings it takes to go around the circle. Estimate what fraction of a string you would need to make the perfect fit if it doesn't come out an even number. Record this number in "Number of Strings" on the data chart.
4. Measure (in degrees) the angle that a single piece of string makes when placed on the circle. Record this angle in "Angle of One String" on the data chart. This is one radian.

5. Now draw a larger circle and repeat this activity. Cut eight pieces of string the length of one radius. Count how many pieces of string it takes to go around, and measure the angle that one piece of string makes.
6. Draw a circle one more time, and make it a bigger or smaller circle than the first two.

Data Chart

Circle	Radius (cm)	Number of Strings	Angle of One String (One Radian)
1			
2			
3			

Post-Lab Questions

1. Radians are normally measured in multiples of π. How many radians are there in a complete circle (360°), measured in multiples of π (i.e., 2π, 5π, 10π)?
2. Did each circle have the same number of degrees per radian, or were they different?
3. How many degrees are there in a radian?
4. In your own words, what is a radian?
5. Arc length is a measure of the distance that an object moves around the outside of a circle measured in meters. If a person knows the radius of the circle and how many degrees were covered, how can he or she calculate the arc length (using what you know about radians)?

LAB 18: CIRCULAR MOTION

This lab is an inquiry activity in that students will likely not know which direction the marble will travel and will discover it with some guidance. It also reinforces the method of preparing a hypothesis before performing an inquiry activity. Although it is OK for students to learn about centripetal force as the force that keeps an object traveling in a circle, and to do centripetal force calculations before this lab, teachers should avoid talking about the path of the object before students perform this activity to keep it from becoming a verification lab.

Topic: Circular Motion
Go to: *www.sciLINKS.org*
Code: THP13

Post-Lab Answers

1. If the object were rotating in a vertical circle, then after the ball leaves the circle, gravity would alter its path and keep it from going in a straight line.
2. "An object in motion tends to stay in motion in a straight line unless acted upon by an external force." The edge of the plate acts on the marble, keeping it rotating. When that force is removed, the marble travels in a straight line.
3. If the centripetal force required to keep it traveling in a circle exceeds the force of friction between the tires and the road, the car will travel in a straight line…right into the wall.

LAB 18: CIRCULAR MOTION

QUESTION

What path will an object traveling in a circle follow if the centripetal force is suddenly removed?

SAFETY

Keep marbles out of the reach of young children. When finished with the plate, do not use it for eating; put it back in the kit.

MATERIALS

Small paper or Styrofoam plate with a rim around the edge, small marble (any type)

PROCEDURE

To keep an object traveling in a circular path requires a force to be applied toward the center of the circle. This force is called *centripetal force*. The centripetal force that keeps the moon traveling in a circle is supplied by the gravitational attraction between the Earth and the Moon. When swinging an object tied to the end of a string in a circle, the tension in the string supplies the centripetal force.

In this lab, you will be answering the question, What path will an object traveling in a circle follow if the centripetal force is suddenly removed?

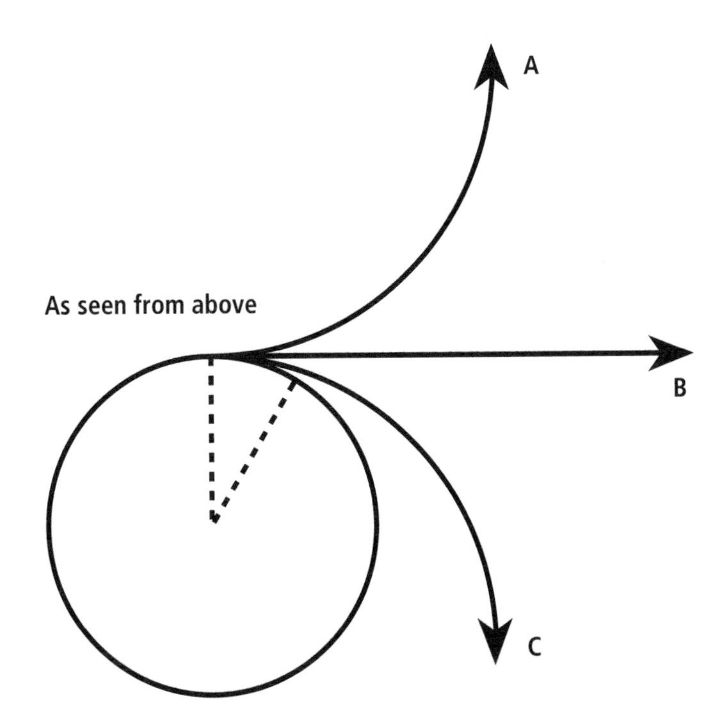

Hypothesis: I think that the object will follow path ____ because ____________
____.

1. Take a Styrofoam or paper plate that has a ridge around the edge and cut out a section of the plate. The section should be shaped like a slice of pie.
2. Put the plate on a flat surface and put a marble in the ridge.
3. Push the marble in the ridge so that it travels around the plate and then out of the removed section.
4. Perform the test several times and record your results.

Post-Lab Questions

1. Explain why this lab would have been more complicated if the object were moving in a vertical circle instead of a horizontal circle.
2. Which of Newton's laws explains the results of this experiment?
3. Use the results of the experiment to explain how a race car might crash if traveling too fast through a turn.

Extension

There is a classic physics demonstration in which a person swings a bucket of water vertically and none of the water comes out at the top. Some explain this with the use of centrifugal force, but centrifugal force is unnecessary in explaining this and is a virtual force that only shows up when the observer is in the bucket's reference frame. Use the results of this lab to explain why the water doesn't fall out if the bucket is being rotated fast enough. Try the experiment if you've never tried it before.

LAB 19: TANGENTIAL SPEED

This lab is purely a thought experiment. Although students are given an introduction to angular speed and tangential speed, they will discover the relationship between the two in this activity. The lab is an inquiry activity in that students do not know this relationship before beginning.

Post-Lab Answers

1. 4π or 12.6 m
2. 12.6 m
3. 12.6 m/5 s = 2.51 m/s and 2π rad/5 s = 1.26 rad/s
4. The tangential speed is numerically double the angular speed.
5. 8π m / 5 s = 5.03 m/s and 2π rad/5 s = 1.26 rad/s
6. The tangential speed is numerically four times the angular speed.
7. Tangential speed = 2 (10π m)/5 s = 12.6 m/s and 1.26 rad/s
8. The tangential speed is numerically 10 times the angular speed.
9. Each time, the tangential speed was R times the angular speed. Therefore the equation would read tangential speed = R(angular speed) or $v = \Omega R$.
10. Answers will vary.

LAB 19: TANGENTIAL SPEED

QUESTION ?

What is the mathematical relationship between angular speed and tangential speed?

Discussion

Angular speed or angular velocity is how many radians an object covers in a certain amount of time. For example, an object that travels one complete circle (2π radians) in 10 s has an angular speed of $2\pi/10$ rad/s, or 0.63 rad/s. The size of the circle makes no difference. An object that travels around a 10 m circle and an object that travels around a 10 cm circle in 10 s both have the same angular speed. The symbol used for angular speed is Ω (omega).

Tangential speed is the same as normal (linear) speed, but for an object that is traveling in a circle. It is calculated the same way as linear speed—distance divided by time. Usually, the distance is the circumference of the circle, so the size of the circle must be known. For example, an object that travels around a circle with a 5 m radius in 10 s has a tangential speed of $(2\pi[5\text{ m}])/(10\text{ s}) = 3.1$ m/s. Notice that angular speed has the units of radians per second and tangential speed has the units of meters per second. You learned in a previous lab that angles measured in radians can appear to not have any units, so sometimes the angular speed will appear to come out as 1/s. We will use the symbol v_t for tangential speed.

Post-Lab Questions

1. If a circle has a radius of 2 m, what is its circumference?
2. If an object traveled all the way around a circle with a radius of 2 m, how far would it travel?
3. If an object traveled all the way around a circle with a radius of 2 m in 5 s, how fast was it moving in meters per second? How fast was it moving in radians per second?
4. From #3, numerically, what was the ratio of the object's speed in meters per second to its speed in radians per second?

5. If an object goes around a circle with a radius of 4 m in 5 s, what is its tangential speed? What is its angular speed?
6. From #5, numerically, what was the ratio of the object's speed in meters per second to its speed in radians per second?
7. If an object goes around a circle with a radius of 10 m in 5 s, what is its tangential speed? What is its angular speed?
8. From #7, numerically, what was the ratio of the object's speed in meters per second to its speed in radians per second?
9. From your answers here, devise an equation that relates angular speed to tangential speed (linear speed).
10. Try your equation and see if it works. An object that goes around a circle with a radius of 2 m with a tangential velocity of 8 m/s has an angular velocity of 4 rad/s. See if your equation gives you the same result.

LAB 20: MOMENT OF INERTIA

This inquiry activity is intended to be an introduction to moment of inertia beyond just mathematical calculations. It provides some background in moment of inertia and allows students to discover that the distribution of mass in the object affects its moment of inertia and therefore its rotation. Soup cans are cylinders, but the thickness of the soup inside them will determine whether they act like a thin-walled hollow cylinder, a thick-walled hollow cylinder, or a solid cylinder.

The equations for the moment of inertia of the three types of cylinders follow:

- Thin-walled hollow cylinder: $I = MR^2$ where M = mass and R = Radius
- Thick-walled hollow cylinder: $I = ½M(R_i^2 + R_o^2)$ where M = mass, R_i = inside radius, and R_o = outside radius
- Solid cylinder: $I = ½MR^2$

An empty soup can would certainly be a thin-walled hollow cylinder. A thick soup like clam chowder would act like a solid cylinder or at least a thick-walled hollow cylinder. A liquid soup like chicken broth would act like a thick-walled hollow cylinder. Concerning cylinders of equal radius and mass, the solid cylinder will have the lowest moment of inertia. Therefore the solid cylinder will win the race down the ramp. The thin-walled hollow cylinder has the highest moment of inertia, so it will lose the race down the ramp but will overtake the other cylinders as they traverse the floor.

Post-Lab Answers, Part 1

1. The thicker soups should win the race down the ramp. The can that won the most races has the lowest moment of inertia.
2. The highest moment of inertia is the can that won the fewest races (thin). The lowest moment of inertia is the can that won the most races (thick).
3. The can with the highest moment of inertia would be least affected by friction and would win the race across the floor.

Post-Lab Answers, Part 2

1. The thin soup can should win as long as the track is long enough.

2. The pattern should be the opposite of Part 1.
3. An object that is easy to spin will be easy to stop spinning. If it rolls down the ramp easily, friction will be able to slow it down easily as well. An object that is difficult to spin will be difficult to stop spinning. If it is difficult to get the object to roll down the ramp, it will be difficult for friction to slow it down.

LAB 20: MOMENT OF INERTIA

QUESTION

Which type of soup can wins a race down a ramp, a can filled with a thin liquid or one filled with a creamy sauce?

SAFETY

Standard safety precautions apply.

MATERIALS

2 cans of soup (one liquid, e.g., chicken broth, and one creamy, e.g., cream of mushroom), a wooden plank to use as a ramp, a smooth floor

PROCEDURE

Up to now, you have dealt mostly with objects traveling in a straight line and ignored rolling or spinning. An object moving in a straight line or at rest is said to have inertia. The more inertia an object has, the more difficult it is to get it started moving or to stop it once it is moving. When an object is rolling, it has a similar "moment of inertia," which is how difficult it is to get the object rolling or to stop it from rolling.

In this lab, you will be racing canned foods down a ramp and across the floor to determine which one has the highest moment of inertia and which one has the lowest. You must recall that an object with a large moment of inertia will be difficult to get started, but once it's rolling, it will be difficult to stop its motion.

Part 1

1. Set up a ramp at least 1 m long with a piece of wood or stiff cardboard. Put it at an angle low enough that a can of soup will not slide down the ramp

but will roll. Make sure that there is a lot of room for objects to roll at the end of the ramp (e.g., across the kitchen or garage floor).

2. Find your two different types of canned food: one that is liquid, like chicken broth, and one that is thick, like gravy. Race the "thin" and "thick" cans down the ramp three times and circle the winner from each race:

   thin/thick **thin/thick** **thin/thick**

   (*Note:* If it is too difficult to race them side by side, you can race them individually and time them with a stopwatch.)

Part 2

In this part, you will be timing how long it takes a can to get from the bottom of the ramp to a certain mark on the floor. The mark should be at least 3 m from the base of the ramp.

With the same ramp as in Part 1, roll each can individually and measure how long it takes for them to travel several meters from the bottom of the ramp. Repeat several times and average the results. Create a data chart from the skeleton chart below and record your data.

thick ________ seconds **thin ________ seconds**
thick ________ seconds **thin ________ seconds**
thick ________ seconds **thin ________ seconds**
average:
thick ________ seconds **thin ________ seconds**

Post-Lab Questions, Part 1

1. Which of the cans won the most races? Does that mean that it has the highest or lowest moment of inertia?
2. Which can had the highest moment of inertia? Which one had the lowest moment of inertia?
3. According to your results, if you were to race the cans across the floor after leaving the ramp, which one would win?

Post-Lab Questions, Part 2

1. Which can had the highest average speed?
2. How did this pattern compare to the pattern from Part 1?
3. Explain in terms of moment of inertia why the patterns came out this way in Parts 1 and 2.

LAB 21: ELLIPTICAL ORBITS

Although not inquiry, this activity is important for students to understand what an ellipse is and what a focus is, and to break misconceptions about Earth's orbit being highly elliptical. This is the perfect place to check to see if students have the misconception that the seasons are caused by the distance between the Earth and the Sun. Ask "What causes the seasons?" and most students will respond that summer is when the Earth is closest to the Sun in its highly elliptical orbit. If they see that the orbit is nearly a perfect circle, they will understand that this cannot be true.

SCI LINKS
THE WORLD'S A CLICK AWAY
Topic: Gravity and Orbiting Objects
Go to: *www.sciLINKS.org*
Code: THP14

Post-Lab Answers

1. The farther apart the thumbtacks, the more eccentric the ellipse.
2. Answers will vary.
3. It means that it is only slightly elliptical, only 0.01% shorter on the minor axis.
4. E is compressed. It is 0.41 in. by 0.42 in. Earth's orbit is not very elliptical.

LAB 21: ELLIPTICAL ORBITS

QUESTION

How elliptical is the Earth's orbit?

SAFETY

Standard safety precautions apply.

MATERIALS

String (kite ~30 cm), 2 thumbtacks, pencil, piece of corrugated cardboard (8.5 × 11 in.)

PROCEDURE

An ellipse looks like a flattened circle. A circle has one center, an ellipse has two foci (plural of *focus*). The eccentricity of an ellipse is a measure of how flattened the circle actually is.

Johannes Kepler determined that the planets orbit the Sun in elliptical orbits. Some of the planets' orbits are far more eccentric than others. The equation for eccentricity is

$$e = \sqrt{1 - \frac{b^2}{a^2}}$$

(where b is the radius in the short direction [semiminor axis] and a is the radius in the long direction [semimajor axis]).

Figure 21.1

Drawing the Ellipse

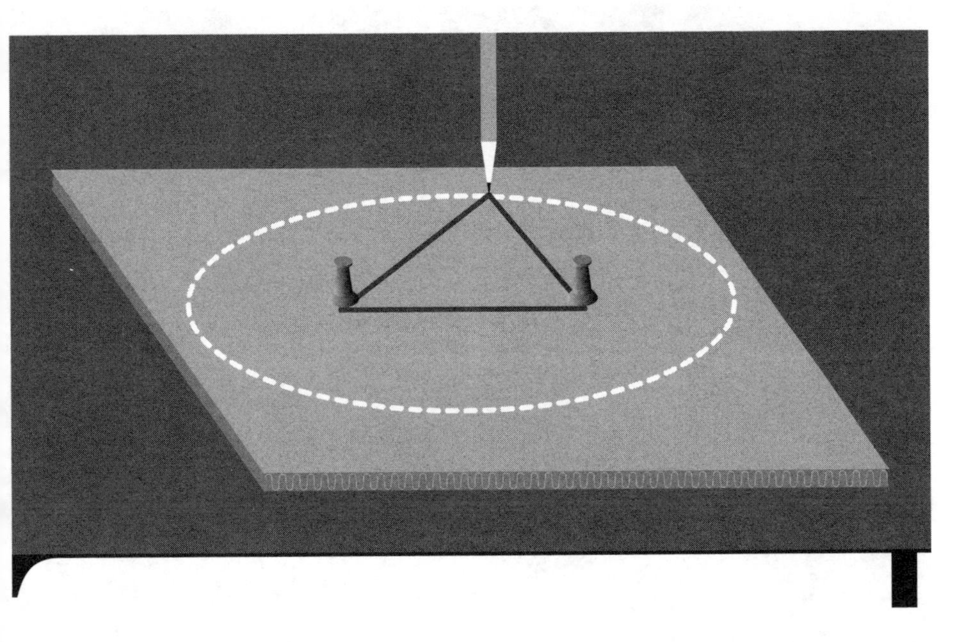

1. Take a piece of string approximately 30 cm long and tie it in a circle. Place a sheet of paper on top of the cardboard and push the two thumbtacks into the paper approximately 4 cm apart. Wrap the string around both thumbtacks.
2. Now use the pencil to draw an ellipse by following the string all the way around the thumbtacks (see Figure 21.1).
3. Remove the thumbtacks and use the holes that they left behind to draw a line along the longest axis of the ellipse.
4. Measure the exact distance between the thumbtack holes and mark the center between them. Draw a line perpendicular to the first line halfway between the thumbtacks (see Figure 21.2, p. 86).
5. Measure the lengths of the two perpendicular lines. The short one is called the "minor axis" and the long one is called the "major axis."
6. Repeat the entire procedure with the tacks closer together.
7. Repeat the entire procedure with the tacks farther apart.

Post-Lab Questions

1. What effect did moving the thumbtacks have on the shape of the ellipse?
2. Calculate the eccentricity of each of the three ellipses using this equation:

$$e = \sqrt{1 - \frac{b^2}{a^2}}$$

Figure 21.2

Ellipse With Axes Drawn

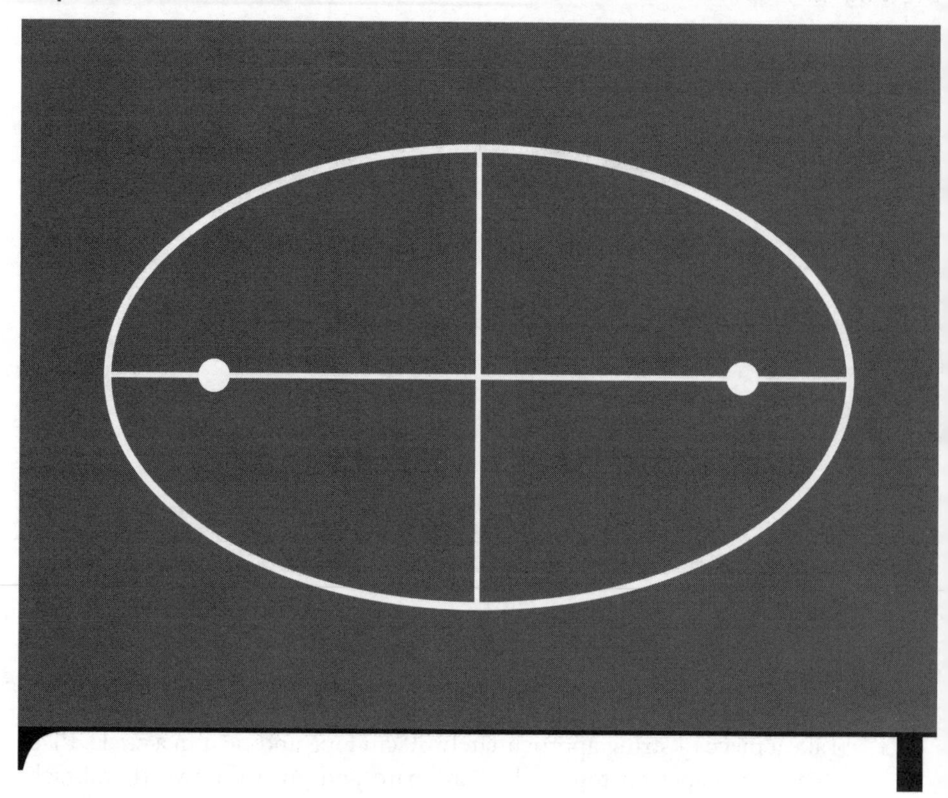

3. The eccentricity of Earth's orbit is 0.017. What does that mean about the shape of Earth's orbit?
4. Earth's orbit is compressed about 0.014%. One of the circles below has been compressed 1.4% (100 times *more* compressed than Earth's orbit). Can you find it? Is Earth's orbit very elliptical?

A 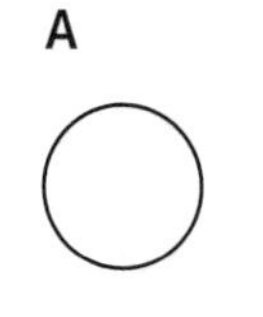 B 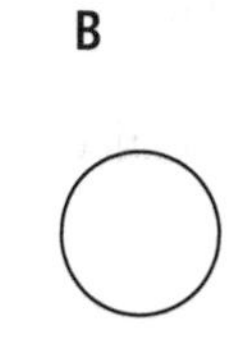 C 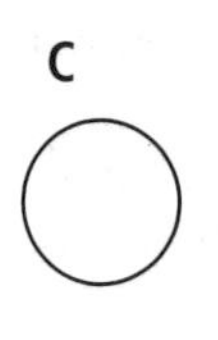 D 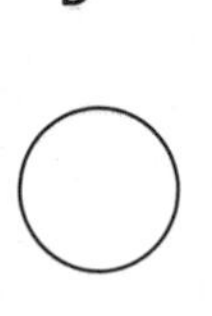 E 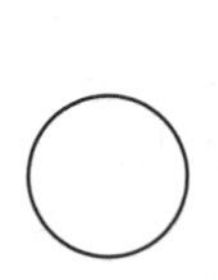 F 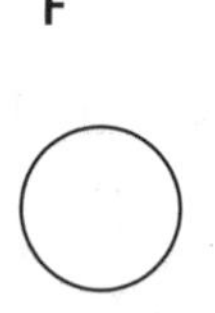 G 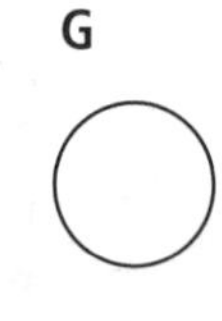

Extension

Many people think that it is the elliptical shape of Earth's orbit that causes the seasons. This doesn't make sense considering that when it is summer in the Northern Hemisphere, it is winter in the Southern Hemisphere. You also learned that Earth's orbit is not very elliptical. Do some research and find out what really causes the seasons. Explain your findings to a friend or family member.

LAB 22: HYDRODYNAMICS

This lab is an inquiry activity in that students do not know the answer and have probably never experienced this phenomenon. Most students have seen that water in U-tubes is always equal height on both sides. This lab will be both a discrepant event as well as an inquiry activity for them. It will also help students when pistons are discussed in class.

Post-Lab Answers

1. The level of fluid changes until the pressure on either side is equal.
2. The less dense side (the side with lower salinity) will be higher because it will take more volume to generate the same pressure. The Pacific side of the canal is on average 20 cm higher than the Atlantic side.

LAB 22: HYDRODYNAMICS

QUESTION

How does the density of liquids affect their interaction when they come in contact with each other?

SAFETY

Do not eat or drink any of the materials when finished.

MATERIALS

2 straws, tape (Scotch or other), water, cooking oil, salt, marking pen

PROCEDURE

Hydrodynamics is the study of the movement of fluids. When designing dams, canals, reservoirs, and dry docks, it's important for engineers to understand how water acts when two bodies of water come into contact. In this lab, you will be simulating a situation like this by connecting two flexible straws. When a freshwater stream flows into the ocean or when the Pacific Ocean meets the Atlantic, scientists need to know how the water will act in order to ensure that it is safe for boats, scuba divers, and animals.

1. Connect the short ends of the flexible straws by sliding one inside the other. Wrap some tape around the connection to seal it watertight.
2. Hold the straws in the shape of a "U" and fill the straws with water so that each side is about ⅔ full. Mark one side L and one side R.
3. Hold the straws level with each other and note the height of the water in each side.
   The water level in the left side is _____ (higher than, lower than, the same as) the right side.

4. Now raise the left straw higher than the right straw and note the height of the water in each side.
   The water level in the left side is _____ (higher than, lower than, the same as) the right side.
5. Now get the two sides even again and put a finger over the end of the left straw and raise the right one. Note the height of the water in each side.
   The water level in the left side is _____ (higher than, lower than, the same as) the right side.
6. Now get the two sides even again and put a finger over the end of the right straw and raise the right one. Note the height of the water in each side.
   The water level in the left side is _____ (higher than, lower than, the same as) the right side.
7. Now get the two sides level again and put some oil in the right side. Note the level of water and oil in each side.
   The water level in the left side is _____ (higher than, lower than, the same as) the right side.
8. Now have an assistant sprinkle some salt in the left straw and allow it time to dissolve. Note the difference between the levels now and before the salt was added.

Post-Lab Questions

1. From your experience in this lab, would you say that the level of fluid on each end of a canal will change until the pressure, volume, or depth on each side is the same?
2. The Panama Canal connects the Pacific Ocean and the Atlantic Ocean. One body of water is saltier than the other. Which side of the canal will be higher, the saltier or the less salty side? (Hint: A salt molecule is heavier than a water molecule.)

Extension

Find the combination of liquids that gives you the biggest difference in the heights of the two sides. Try liquids that you have around the house, such as rubbing alcohol, Kool Aid, etc.

SECTION 2:
Forces and Energy

LAB 23: UNBALANCED FORCES

This activity is one for which you can find a lot of different explanations on the internet, but most of them are wrong. Some say that the soap is ejected from the back of the boat. Others say that water is ejected from the back of the boat. Still others say that there is surface tension in front of the boat and none in back of the boat. The one that is most accepted by scientists is the surface tension explanation. That is the one that we will use here.

This activity is inquiry because students have likely not done the activity before, you have not talked about unbalanced forces before, and students likely cannot guess what will happen before trying the experiment. Before completing this activity, you should teach students how to draw free-body diagrams.

See the next page for boat outlines. Either print the page onto printable transparency sheets, or photocopy the page onto transparency sheets meant for copiers. Do not put regular transparencies into a laser printer or copy machine; they will melt and damage the machine. There are special transparency sheets made for this purpose. Laser printer transparencies and dry toner copier transparencies are the same, but write-on transparencies and inkjet transparencies will destroy a laser printer or copy machine.

Cut out a set of three boats for each student. You can get 21 sets from one sheet of transparency. The students can do the fine cutting of each boat; you simply cut a rectangle around three boats.

Topic: Balanced and Unbalanced Forces
Go to: *www.sciLINKS.org*
Code: THP15

Post-Lab Answers

1. Before the soap, the boat has surface tension pulling it in all directions, and it does not move. After the soap, there is more surface tension pulling forward than backward, so the boat moves forward.
2. Slightly. The boat moves diagonally when the soap comes out of one side or the other. This suggests that the surface tension explanation is correct. It would spin in circles if it was ejecting material from the hole.
3. The force is very small, but in a very still lake, it would likely move a small boat slightly. It would only work once, though. Once the soap spread around the lake, it would not work for anyone else's boat, so it would not be a practical way to move boats around a lake.

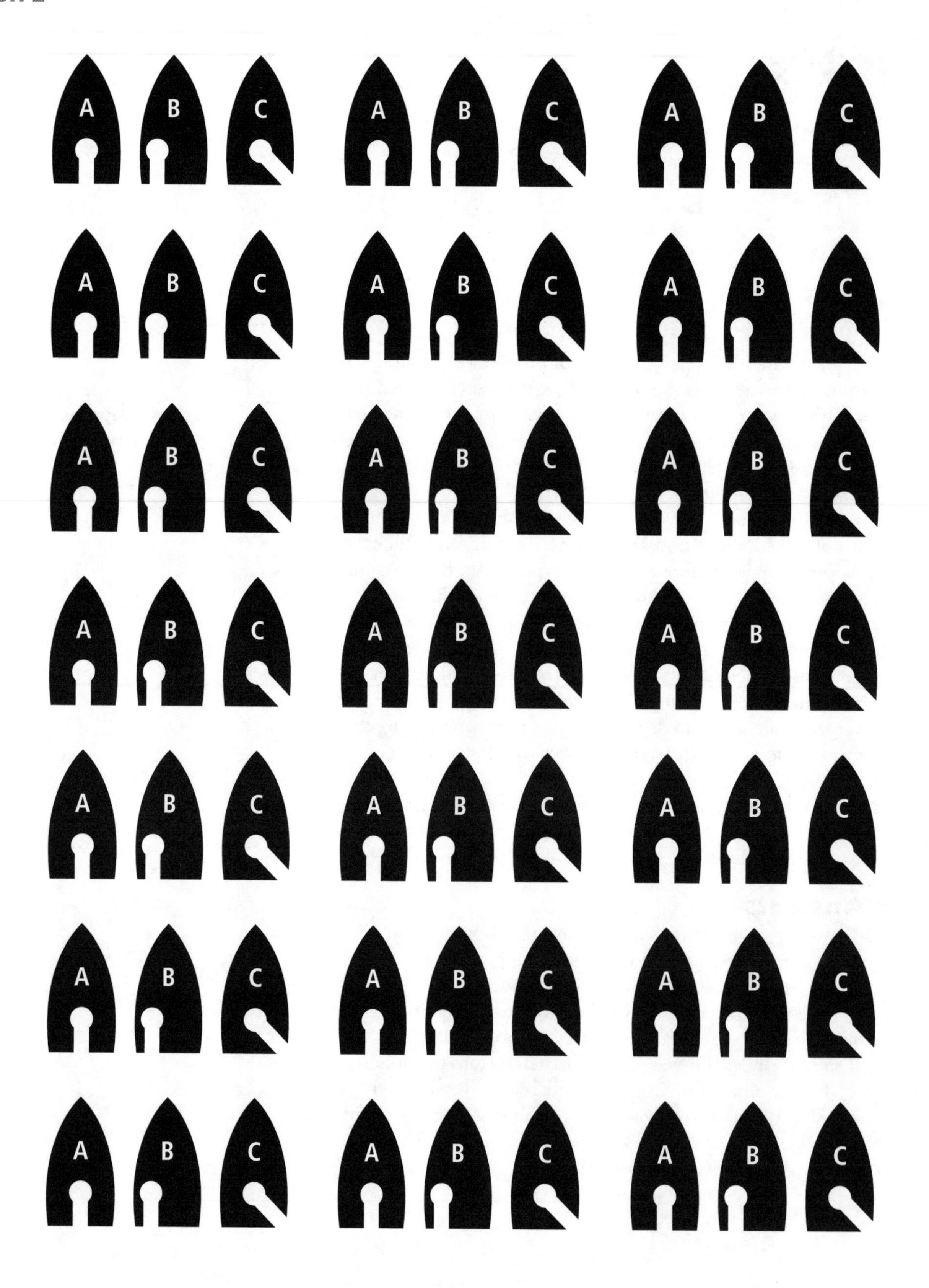
A
B
C
A
B
C
A
B
C

LAB 23: UNBALANCED FORCES

QUESTION

What happens to an object that has unbalanced forces acting on it?

SAFETY

Be careful with scissors.

MATERIALS

Transparency sheet with boat drawings, liquid soap, large container of water (such as your materials box), scissors

PROCEDURE

To get an object to move, there must be an imbalance in the forces acting on it. For example, if one person pushes a car forward with a force of 100 N and another person pushes it backward with a force of 100 N, it won't move. But if both people push the car forward each with a force of 100 N, then it will move. Physics also tells us that if the force is not acting through the center of mass, then a torque is applied and the object may rotate.

In this lab, you will be using surface tension to create uneven forces to move a small plastic boat. Water has a property called *surface tension* (technically, it's called *cohesion* if it's pulling other water molecules, or *adhesion* if it's pulling other nonwater molecules). Surface tension is the name of the force created by water sticking to other water molecules and other substances. It is the reason that water's surface is curved in a thin cylinder (called a *meniscus*) as in Figure 23.1, page 96. It pulls on everything that is put into the water.

Soap has the ability to break surface tension, and it will be used to create the uneven forces on the boat.

Figure 23.1

Water in a Thin Cylinder

1. Fill a large container with tap water and let it sit until it is motionless.
2. Cut out the three boats.
3. You will be putting a drop of dishwashing liquid or liquid soap in the center of each boat. This will break up the surface tension in the rear of the boat. Draw a free-body diagram* and a sketch of what you think the motion of each of the boats will be. Some forces that you should include are weight, buoyancy, and surface tension.
4. Now put the first boat in the middle of the bowl of water. Dip a small straw in the liquid soap and put a drop of soap in the hole in the middle of the boat.
5. Completely rinse all of the soapy water out of the bowl and repeat Step 4 for the second and third boats.
6. Draw a diagram of the motion of the boats.

*A free-body diagram is a diagram of the object with arrows representing all the forces on it. The longer the arrow, the larger the force. The direction of the arrow points in the direction of the force.

Post-Lab Questions

1. Why does the boat not move when you first put it in the water? Why does it move after you add the soap?
2. Did the location of the ridge coming out of the back of the boat have an effect on the motion of the boat?
3. Do you think this could work with a small boat in a lake? Explain your answer.

Extensions

There is a classic demonstration involving surface tension that is easy to perform and now should be easy to explain. Get a bowl of water and sprinkle some black pepper on the surface. Put a single drop of dishwashing liquid in the center of the bowl and observe. Can you explain this behavior?

There are bugs that use surface tension to walk on water. Some of them use chemicals to propel themselves through the water by breaking up the surface tension behind them. Use the internet to research these creatures and try to find video of the process.

LAB 24: CENTER OF MASS 1

This activity will help students see that a low center of mass helps objects balance. This is important engineering in car and motorcycle racing and robotics. This activity is inquiry in that students would probably predict that you could not get one object completely over the side of the table before they begin. Question 3 is Level 2 inquiry because the student is given the question and comes up with his or her own procedure and conclusions.

SCI LINKS. THE WORLD'S A CLICK AWAY
Topic: Center of Mass
Go to: *www.sciLINKS.org*
Code: THP16

Post-Lab Answers

Part A

1. The closer together the bottom dominoes are, the more mass there is near the bottom. This lowers the center of mass, which allows you to spread out the top dominoes and get the top one off the edge of the table.
2. The stack acts like a series of structures. That is, the top domino must balance on the second domino. The top two must balance on the third, and so on. The stack must not break anywhere. As long as the whole stack follows this rule, the top domino can actually be as far as one wishes (given enough dominoes and enough patience).
3. Answers will vary based on the density and length of the domino. Anything less than 5 is probably impossible unless they are unique dominoes.

Part B

1. Most people have a center of mass that is slightly above their waist. When we lean forward, our center of mass moves forward also. Normally, our lower bodies move backward to keep the center of mass above the feet. When standing against a wall, this is not possible, so the new center of mass goes beyond the feet and you fall over.
2. When kneeling, you have a larger base, so you can lean farther, but eventually, the center of mass goes beyond the knees and you fall.
3. In general, females have a lower center of mass.

LAB 24: CENTER OF MASS 1

QUESTION

How are center of mass and balance related to each other?

SAFETY

Be careful when laying on the edge of a bed or stair rail. Do not attempt this if the surface is more than 2 ft. high.

MATERIALS

8–10 dominoes, coins, or small blocks of wood

PROCEDURE

Part A

1. Begin by placing a domino near the edge of the table without hanging over the edge.
2. Now stack another domino on top but overlapping a little.
3. Put more dominoes on the stack, overlapping each one by a little more (see Figure 24.1, p. 101).
4. When you get to the top, the top domino should be completely over the edge of the table. If it is not, play with the positions of the dominoes until the top domino is over the edge.

Part B

1. Find your center of mass by laying on the edge of a bed or the bottom of a stair handrail until you balance, and note the place where the railing is touching your body.

Figure 24.1

Domino Stack on Table

2. Sketch a diagram of yourself (it can be a stick figure) and show where your center of mass is compared to your waist.
3. Stand with the back of your feet against a wall. Without moving your feet, try to bend down and touch your toes.
4. Now kneel down on the ground and put a pencil about 75 cm in front of you. With your hands behind your back, try to lean over and pick up the pencil with your teeth.

Post-Lab Questions

Part A

1. Why is it important that the bottom dominoes do not overlap by very much? Consider center of mass of the stack in your answer.
2. Why does the top domino not fall off even though it is over the edge of the table?
3. Perform another experiment to determine what the minimum number of dominoes is to get the top one completely extended over the edge of the table.

Part B

1. Using the idea of center of mass, explain why you fell over when you tried to touch your toes.
2. Explain why you couldn't pick up the pencil with your teeth.
3. Compare where center of mass with your waist is in comparison to at least five other people (a mixture of males and females) and see if you can make a statement about whose center of mass seems to be farther away from their waist.

Extension

Repeat Part B, Step 3 but with a short chair in front of you. Try to lean over and stand straight up with the chair in your hands. Have several other people do this and you might find that one gender is more successful at this activity than the other. Use your answer to Post-Lab Question 3 to explain this.

LAB 25: CENTER OF MASS 2

This activity is inquiry because students have likely never used this method for determining center of mass. The activity is also useful when building soda bottle rockets for estimating the center of pressure. Students can make a cardboard cutout of the profile of their rocket and determine its center of mass, which will be the center of pressure of the rocket.

On the next page is an image of the state of California for you to copy and distribute to students.

Post-Lab Answers

1. The centers of mass in Steps 3 and 4 should be pretty close. The balance point in Step 4 should be within the circle in Step 3 if students drew their lines carefully. How close the center of mass in Step 2 is depends on students' prior understanding of center of mass.
2. Adding weight to an object moves the center of mass from the prior point in a direction pointing directly toward the object. The mass of the object and distance from the original center of mass determine how far the center of mass moves.
3. Removing weight from an object moves the center of mass from the prior point in a direction pointing directly away from the hole. The size of the hole determines how far the center of mass moves.

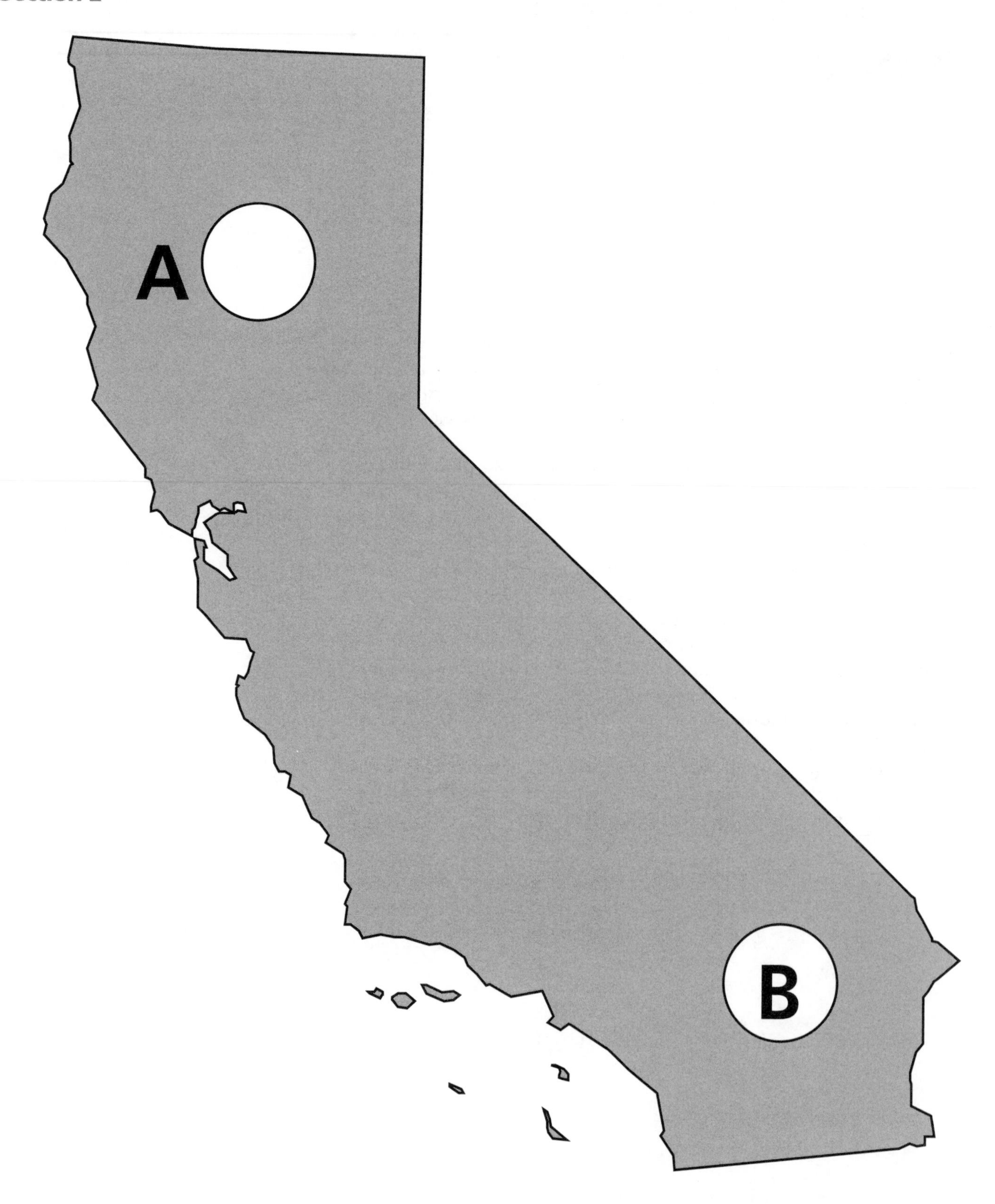
A
B

LESSON 25: CENTER OF MASS 2

QUESTION

How can a plumb bob be used to determine the center of mass of an irregular object?

SAFETY

Standard safety precautions apply.

MATERIALS

Cardboard (8.5 × 11 in.), scissors, thread, washer

PROCEDURE

One way to find the center of mass of an object involves a plumb bob. A plumb bob is simply a weight hanging from a string. Because gravity pulls straight down, the string always hangs straight down. This is useful in building things like door frames that are perfectly square.

In this lab, you will be cutting out the shape of the state of California and finding its center of mass.

1. Cut out the outline of California and tape it to a piece of cardboard, then cut the shape out of cardboard.
2. Estimate where you think the center of mass will be and put an *X* there.
3. Now hold the state from an edge by lightly pinching it between a finger and thumb. Let it rotate freely to find its balance point. Tie a washer to the end of a 30 cm piece of thread. Put the weighted thread between the same finger and thumb, and allow it to dangle as well.
4. Trace the string on the cardboard. Repeat three more times (see Figure 25.1, p. 106). Put a circle where the lines meet. If the lines don't meet perfectly, make the circle big enough to encompass them all.

Figure 25.1

Rotate your object around several times and draw a line along the string.

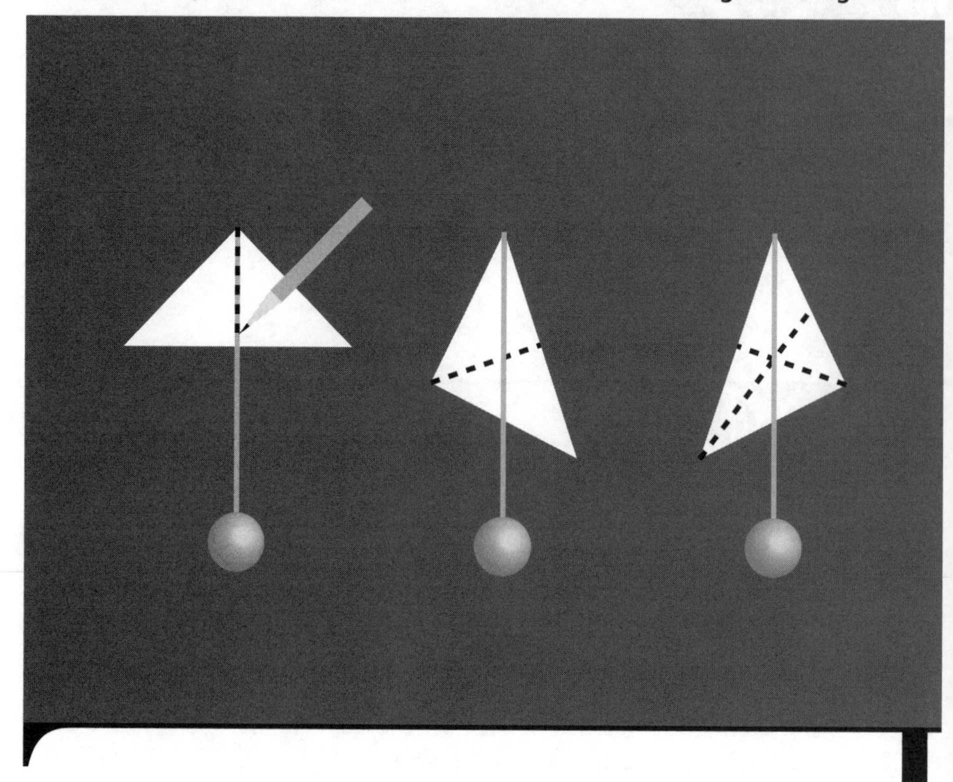

5. Now balance the state on the tip of a pencil and mark the location of the balance point with a ★.
6. Now tape a penny to the back of the state at point A and balance it to find the center of mass and mark it with a ☺.
7. Now cut a circle out of the state at point B and balance it to find the center of mass and mark it with a ⬢.
8. Turn in your cardboard cutout with the marks on it.

Post-Lab Questions

1. Were your three centers of mass pretty close in Steps 2, 3, and 4? If not, which do you think was the most accurate?
2. Write a rule explaining how adding weight to an object affects its center of mass.

   Rule: When you add weight to an object, its center of mass moves ______________________.
3. Write a rule explaining how taking weight off of an object affects its center of mass.

   Rule 2: When you take weight off of an object, its center of mass moves _________________.

LAB 26: CENTER OF MASS 3

This lab is an inquiry activity if it is done before the discussion of how center of mass and balance are related. However, students should have an idea of what center of mass is before they start the lab. Students should see that when the center of mass is high (the can is full), the can will tip over. The more fluid in the can, the farther to the right the center of mass will go on the diagram. When the water level gets lower, it centers itself in the V-shape of the bottom of the tilted can and the center of mass remains above the base over quite a range.

If students go below the 20% mark, they may be surprised that the can still tips toward the top, but it is the shape of the can that causes this when the water is less of a factor. In the diagram, there is far more can to the right of the base than there is to the left.

Post-Lab Answers

1. It will be directly above the base (the part of the can that touches the table). The height cannot be determined because there is a lot of empty soda can near the top that affects the balance point. If the can were weightless, it would still be difficult to determine because of the shape of the can when it is sitting at an angle.
2. You would have to consider how much salt is in the shaker. If there is too much or too little, the center of mass cannot be above the balance point. Adding the salt lets the shaker lean on some salt crystals, which increases the size of the base, making it easier to balance.
3. They should add weight to the bottom to keep it from toppling over by moving the center of mass down. In fact, engineers have already done this. The other possibility would be to increase the size of the base. This could be done by adding supports or widening the base structure.

LAB 26: CENTER OF MASS 3

QUESTION

Where must the center of mass be in relation to the base of an object in order for the object to stay upright?

SAFETY

Opened soda cans may have sharp edges; be careful not to cut yourself.

MATERIALS

Unopened soda can, ruler, tape (Scotch or other)

PROCEDURE

The center of mass is similar to the point at which an object will balance on your finger. The only difference is that center of mass is three-dimensional and is inside an object, not on the outside. For example, a symmetrical meterstick that is 1 m long, 3 cm wide, and 0.50 cm thick has a center of mass at 50 cm along its length, 1.5 cm along its width, and 0.25 cm along its thickness. Sometimes the center of mass is at a location where nothing is there. The center of mass of a paper clip is such that it is located in an empty space just off center.

Center of mass is important in physics because it is the center of mass of an object that follows a parabola when in projectile motion, and the location of an object's center of mass determines whether it will be stable or fall over.

1. Take an unopened can of soda and try to balance it at an angle without the help of any other objects. Use the ridge at the bottom of the can to help. Mark with a piece of tape approximately where you think the center of mass is when the can is tilted at an angle. Keep in mind that there is air at the top. Try to balance the can on your finger there to check. With liquid

inside, it may not stay, but you can tell if you are close. Record your results and draw your diagram in a data chart (p. 110).

2. Now open the can and remove approximately 20% of the liquid and try to balance it on the ridge again. Which way did it try to fall, toward the top of the can or the bottom? Draw a diagram to explain why. Include the approximate center of mass in your diagram (see Figure 26.1). Remember that as you take liquid out of the can, the center of mass changes.

Figure 26.1

Diagram With Center of Mass Marked

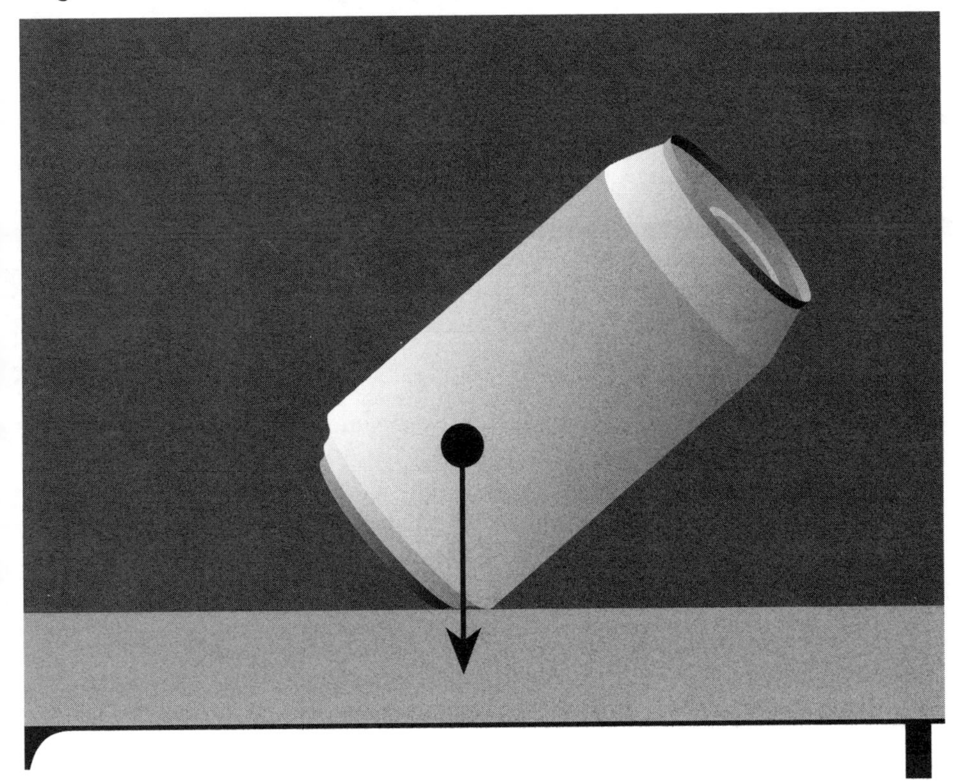

3. Remove another 20% of the liquid and try to balance it on the ridge again. Which way did the can fall? Draw a diagram.
4. Keep removing 20% of the contents of the can until it is empty. Note when it does balance.
5. Repeat this experiment while filling the can with water. Find the range of water levels between which you can balance the can on the ridge. Mark them on the can and measure how far they are from the bottom of the can.

Data Chart

Percent	Balance? Yes/No	Draw a diagram with center of mass marked (similar to Figure 26.1).
100		
80		
60		
40		
20		

Post-Lab Questions

1. When the can is balanced at an angle, where is the center of mass?
2. Some people perform a restaurant trick by sprinkling some salt on the table and then balancing the saltshaker at an angle in the sprinkled salt. What are some things that you would have to consider before trying this? Why would you sprinkle salt on the table?
3. According to the results of this experiment, if engineers were to try to stabilize the Leaning Tower of Pisa, should they add weight to the top, middle, or bottom of the tower to keep it from falling over? Explain why.

Extension

Play a trick on your family and see if they can figure it out. Using the marks on your can in this experiment, poke a hole with a thumbtack in an unopened can of soda between the two lines and let the soda come out until the liquid is at that level. Do this over a sink so as not to make a mess; the soda shoots out very fast. Now balance the soda can and let your family try it with normal cans of soda without a hole. They won't be able to do it. Put a small piece of tape over the hole to ensure that the soda won't come out unexpectedly.

LAB 27: SPRING CONSTANTS

This inquiry activity should be completed before students have learned about spring constants. It should be pointed out to students that if a rubber band or spring is stretched too much, the spring constant is not constant.

Graphs will vary based on the stiffness of the rubber band and the weight of the washers. Graphs should be very close to a straight line.

Post-Lab Answers

1. Answers will vary. Teachers should determine the approximate spring constant before handing out the rubber bands and allow some degree of error to account for the age and temperature of the rubber bands.
2. Answers will vary, but students should extrapolate from their graphs to determine the answer. If they use the spring constant and Hooke's Law to calculate it, that is also acceptable.
3. Yes, the rubber band gets very stiff near its elastic limit and eventually will break.

LAB 27: SPRING CONSTANTS

QUESTION

How much does a rubber band stretch with different weights applied?

SAFETY

Standard safety precautions apply.

MATERIALS

Rubber band, metric ruler, 6 large washers, paper clip

PROCEDURE

A spring constant is a way to measure how strong a spring is. A thin rubber band will have a low spring constant and a garage door spring will have a very high constant. The symbol for the spring constant is k and it shows up in Hooke's Law, $F = -k\Delta x$. This equation is used to figure out how much a spring will stretch if a certain amount of force is applied to it. F stands for force or weight measured in Newtons, k is the spring constant measured in N/m, and Δx is the amount measured in meters that the spring stretches when the weight is hung from it. The negative sign is present because the weight and the spring act in opposite directions. If a spring is stretched, it pulls back. If a spring is squashed, it pushes back.

1. Attach a bent paper clip to the end of the rubber band to be used to hold the weights. Find out from your teacher the mass of one of the washers. Hang one washer from the rubber band just to straighten it out and measure the rubber band's length.

   Length _______ m

2. Now put three washers on the rubber band and measure its length.

   Length _______ m

3. Now put six washers on the rubber band and measure its length.

   Length _______ m

4. Ask your teacher for the mass of a washer.

   Mass _______ g = _______ kg

5. Create a graph that indicates how much the rubber band stretched from its original length on the horizontal axis and the weight of the washers (not the mass) on the vertical axis. Make the horizontal axis stretch far enough to accommodate 10 washers so that you can extrapolate information from it. *Note:* Recall that to turn mass to weight, use the formula $W = mg$

   KEEP YOUR CALIBRATED RUBBER BAND, AS YOU MAY NEED TO USE IT TO MAKE MEASUREMENTS IN OTHER LABS. ALSO KEEP A COPY OF YOUR GRAPH SO THAT YOU MAY USE YOUR CALIBRATED RUBBER BAND TO WEIGH OBJECTS IN THE FUTURE.

Post-Lab Questions

1. When rearranging Hooke's Law to solve for k, you get $F/\Delta x$. So the slope of your graph is the spring constant of your rubber band. Determine the slope of your graph. If it is not a straight line, ignore the point with only one washer and use the points with three and six washers instead. Two points always make a straight line.
2. How long would the rubber band have been if you were to hang 10 washers from it?
3. From your experience, is there a limit to how many washers you could hang before this pattern falters? Do a mini-experiment to check your answer.

LAB 28: SPRING COMBINATIONS

Although this would not be considered a standard lab, it does help students with projects and other labs that involve the combinations of springs or rubber bands. Even if the activity is not used elsewhere, it allows students one more chance to practice using Hooke's Law and reinforces the idea of spring constants. It is also very useful in projects such as egg drops, robots, slingshot launchers, and catapults.

Post-Lab Answers

1. No, adding string should not change the elasticity of the rubber band.
2. No, adding more rubber bands in line does not change the length of any of the rubber bands. The tension in the entire string stays the same no matter how many bands are added.
3. Yes, adding more rubber bands side by side does lessen the stretching of each rubber band because they are helping hold the weight.

LAB 28: SPRING COMBINATIONS

QUESTION

Is the spring constant of a combination of rubber bands the sum of the individual spring constants?

SAFETY

Do not leave rubber bands or plastic bags where young children might choke on them.

MATERIALS

2 identical rubber bands (one calibrated—created in Lab 27), thread, resealable snack- or sandwich-size bag, water

PROCEDURE

In this lab, you will see how combining springs in different configurations affects the reading on your spring scale. You will be combining rubber bands and strings and noting the result. You will also see how adding other material between the rubber band and the mass affects the reading.

1. Put approximately 100 ml of water in a resealable bag and put a paper clip through the corner from which to hang the bag. It does not have to be exactly 100 ml.
2. Hang the bag from your calibrated rubber band and measure the mark.

   ___________ cm

3. Now put 5 cm of thread between the rubber band and the bag of water and measure again.

   ___________ cm

4. Put 10 cm of thread between the rubber band and the bag of water and measure again.

   ___________ cm

5. Predict what will happen if you put another rubber band between the calibrated rubber band and the bag of water.

   ____

   ____

6. Now put another rubber band between the calibrated rubber band and the bag of water and measure the calibrated mark.

   ___________ cm

7. Now add another rubber band between the two rubber bands from before and the bag of water and measure again.

   ___________ cm

8. Connect one rubber band to each end of the 10 cm piece of thread and drape it over a smooth, round pencil. Hang the bag of water from the calibrated rubber band and hold the other rubber band still. Measure the calibrated mark.

   ___________ cm

9. Now put the two rubber bands side by side and hang the bag of water from both of them and measure the calibrated mark.

   ___________ cm

Post-Lab Questions

1. Did adding thread to the configuration change the length of the calibrated mark?
2. Did adding more rubber bands in-line (in series) with the calibrated rubber band change the length of the calibrated mark (Steps 6 and 7)?
3. Did adding more rubber bands side by side (in parallel) with the calibrated rubber band change the length of the calibrated mark (Step 9)?

LAB 29: CENTRIPETAL FORCE

This activity is inquiry because students should not yet know the equation for centripetal force. Although they will not exactly derive the equation from this lab, they will determine that there is a direct relationship between force, mass, and velocity, and an inverse relationship between force and circumference (radius). They will not be able to derive the square relationship with velocity.

If the teacher feels that swinging washers is too dangerous, students may be given two rubber stoppers or something softer than a washer.

Topic: Centripetal Force
Go to: *www.sciLINKS.org*
Code: THP17

Post-Lab Answers

1. Increase the radius of the turn, decrease the speed, and increase the friction between the tires and the road (the centripetal force). "Decrease the mass" will, in fact, not help because the mass of the car contributes to friction. Reducing the mass will reduce the centripetal force required to turn, but it will also reduce the friction between the tires and the road.
2. Centripetal force and mass, direct. Centripetal force and velocity, direct. Centripetal force and radius, indirect.
3. Yes, it should.

LAB 29: CENTRIPETAL FORCE

QUESTION

What factors affect centripetal force?

SAFETY

Perform this activity outside to avoid hitting anything in your home.

MATERIALS

Rubber band, 4 large washers, 1 paper clip, thread, stopwatch

PROCEDURE

Newton's first law says that an object in motion will stay in motion in a straight line until a force acts on it. This implies that to make something move in a circle requires a force acting on it. For planets orbiting the Sun, that force is gravity. For a tetherball swinging around a pole, the force is the tension in the string. The force that is required to keep something moving in a circle is called *centripetal force*. In this lab, you will be discovering how mass, radius of the circle, and speed affect centripetal force.

Remember the rules of constants and variables when performing these procedures. You can only change one of the three variables at a time to see their effect.

1. Tie a thread of approximately 30 cm to the calibrated rubber band. Tie a paper clip to the other end of the string. Put one washer on the paper clip.
2. Wrap the rubber band around a finger on your left hand. Hold your left hand over your head. Hold the washer in your right hand and toss it so that you can spin it over your head (reverse if you are left handed). Wet

your finger if the rubber band sticks to your finger. Practice this until you can do it smoothly.

3. Have a helper start the stopwatch and count off the seconds to you (one, two, three, four, five…). Synchronize the spinning of the washer so that it comes around once per second. Observe the length of the rubber band.
4. Add three more washers and repeat Step 3. Was the rubber band longer or shorter?

   Longer (circle one) Shorter

5. Finish this sentence:

   The larger the mass of the object, the __________ (more/less) centripetal force it takes to keep it moving in a circle.

6. Put two washers on the paper clip and get the washer spinning once per second. Again, observe the length of the rubber band.
7. Now get the washer going around twice per second. Was the rubber band longer or shorter?

   Longer (circle one) Shorter

8. Finish this sentence:

   The faster the object is moving, the __________ (more/less) centripetal force it takes to keep it moving in a circle.

9. With two washers on the paper clip, get the washer spinning once per second. Again, observe the length of the rubber band.
10. Now put a 60 cm string on the rubber band with two washers on a paper clip on the end and get it spinning once every two seconds. Observe the length of the rubber band. Was the rubber band longer or shorter? (*Note:* You are spinning it once every two seconds in order to keep the tangential velocity of the washers equal to the velocity of the washers in Step 9. Because the circle is twice as large and the object is going around in twice the time, the tangential velocity will be the same.)

    Longer (circle one) Shorter

11. Finish this sentence:

    The larger the radius of the circle on which an object is traveling, the __________ (more/less) centripetal force it takes to keep it moving in a circle.

Post-Lab Questions

1. A car that is turning will skid when the centripetal force required to make it turn is greater than the force of friction between the tires and the road. What are two things that a driver could do to decrease the chance that he or she will skid on a turn, based on what you learned in this lab?
2. A direct relationship between two variables means that when one goes up, the other goes up, and when one goes down, the other goes down. An

inverse relationship means that when one goes up, the other goes down, and vice versa. Was the relationship between centripetal force and mass direct or inverse? The relationship between centripetal force and velocity? The relationship between centripetal force and the radius of the circle?

3. The equation for centripetal force is $F = mv^2/r$. Does this agree with your answers to question 2?

LAB 30: CONSERVATION OF ENERGY

This activity is inquiry because students have not yet been exposed to the idea of gravitational potential energy being converted into kinetic energy. Students should be able to measure this by performing this activity. They will compare the velocity that they calculate using conservation of energy to a distance/time calculation that they learned previously.

Topic: Conservation of Energy
Go to: *www.sciLINKS.org*
Code: THP18

Post-Lab Answers

1. There should be very little difference between the two. Teachers must explain to students what is meant by "very little difference." Some students will get 0.210 and 0.208 and think that this is a huge difference. This can lead them to the wrong conclusion in this lab.
2. Answers will vary, but the calculation involves setting the equation for potential energy equal to the equation for kinetic energy and solving for velocity. The masses cancel out of this calculation, so the mass of the marble doesn't really matter. $gh = ½v^2$
3. Answers will vary. Possible sources of error include difficulty synchronizing the dropping of the marble with the starting of the stopwatch, difficulty in timing such a short interval accurately, and difficulty in dropping the marble from exactly the correct height.

LAB 30: CONSERVATION OF ENERGY

QUESTION

How do you use conservation of energy to determine the final velocity of a falling object?

SAFETY

Do not leave the marble in a place where someone might step on it or a child might swallow it.

MATERIALS

Marble (any), stopwatch, ruler, pie pan

PROCEDURE

When an object is held above the ground, it has a potential energy relative to the ground that depends on its mass and its height above the ground. When the object is dropped, the potential energy converts to kinetic energy, which depends on mass and velocity. Just before the object hits the ground, all the potential energy is converted to kinetic energy.

Because all the energy is converted, the potential energy at the beginning is equal to the kinetic energy at the end because the kinetic energy is 0 at the beginning and potential energy is 0 at the end. So, one can combine the two individual equations because they are equal to each other in this case.

$$mgh = \frac{1}{2}mv^2$$

Note: This equation is not generally true for all situations. It only works for a situation exactly like the one in this activity.

The masses cancel out on each side and a person can calculate the velocity of the object just before it hits the ground from just its drop height.

In this lab, you will drop an object from different heights and time it as it falls. You will then calculate the final velocity using the techniques learned in Lab 4: Final Speed and compare the values to the conservation of energy method. Because the only thing that you have to measure in the conservation of energy method is the height, and for the acceleration method you measure height and time, the energy method has less inherent errors in it.

1. Have a partner help you time how long it takes for a marble to drop from 1 m to the floor. Put something on the floor that will make a loud noise to help you in stopping the watch at the correct time; pie pans work well for this.
2. Measure the time three times and record the data in a chart similar to the one below.
3. Repeat with a marble dropped from 2 m.
4. Get the mass of a marble from your teacher. Mass = _____ g = _____ kg

Data Chart

Height	Potential Energy	Time 1	Time 2	Time 3	Average Time
1 m					
2 m					

Calculations

1. What is the potential energy of the marble at 1 m?

   ____________________________ J

2. What is the potential energy of the marble at 2 m?

   ____________________________ J

3. What is the average speed of the marble falling from 1 m?

   ______________________ m/s

4. What is the average speed of the marble falling from 2 m?

   ______________________ m/s

5. What is the final speed of the marble falling from 1 m?

   ______________________ m/s

6. What is the final speed of the marble falling from 2 m?

   ______________________ m/s

7. What is the kinetic energy of the marble falling from 1 m just before it hits the ground?

    ____________________ m/s

8. What is the kinetic energy of the marble falling from 2 m just before it hits the ground?

    ____________________ m/s

Post-Lab Questions

1. What is the percentage difference between the potential energy at the beginning and the kinetic energy at the end of the marble dropped from 1 m? Of the marble dropped from 2 m?
2. Use the potential energy at the beginning to calculate the final velocity at the end.
3. Calculate the percentage difference between the velocity calculated through conservation of energy and through distance/time. Give several possible sources of this difference.

Extensions

Find three toys or games that convert one kind of energy to another. Bring the toys to class and share them with the rest of the class. Find video clips online that demonstrate conservation of energy and have your teacher compile them and show them all to the class.

LAB 31: CONVERSION OF ENERGY

Teachers may want to show students a completed soda can vehicle because the written instructions can be difficult to follow for some students. These cans can travel a long distance (an entire school hallway is possible). Soda cans may have sharp edges, so encourage students to be careful when constructing their vehicles.

Topic: Potential and Kinetic Energy
Go to: *www.sciLINKS.org*
Code: THP19

Post-Lab Answers

1. Answers will vary but usually involve larger rubber bands, modifications to grip the floor, and lubricants to reduce friction between the pencil and the can.
2. When you wind the rubber band, it contains a lot of elastic potential energy. When you let go of the pencil, the potential energy of the rubber band is converted to kinetic energy in the spinning of the pencil. When the pencil hits the ground, the can spins instead, which makes the car move forward.
3. The circumference of a can is approximately 20 cm. For each spin of the pencil, the car should move 20 cm (if slippage is ignored). Fifty rotations of 20 cm will be 1,000 cm or 10 m.

   $C = 2\pi r = 2(3.14)(3.25 \text{ cm}) = 20.4 \text{ cm}$

   $50 \text{ rot.}(20 \text{ cm/rot.}) = 1{,}000 \text{ cm}$

LAB 31: CONVERSION OF ENERGY

QUESTION

How far can a vehicle go by converting elastic potential energy to kinetic energy?

SAFETY

Soda cans may have sharp edges; be careful not to get cut.

MATERIALS

Rubber band, pencil, empty soda can, bead with hole through center (standard marble size), nail

PROCEDURE

You will be making an amazing device that is so simple yet so effective. With just a rubber band, pencil, bead, and aluminum can, you will make a vehicle that can go all the way across a large classroom, sometimes more than once. By winding up the toy, you will be loading it with energy that will be converted into the motion across the floor.

1. Get an empty soda can and leave the tab on the top.
2. Use a nail to carefully punch a hole in the bottom of the can, being sure that the hole is smooth and not sharp.
3. Put one end of the rubber band through the bead and then through the hole in the bottom of the can. Put the pencil through the rubber band to keep it from going inside. Put the rubber band through the can. Use an open paper clip or tweezers to pull the other end of the rubber band through the mouth of the can and wrap it around the tab at the top of the can (see Figure 31.1).

Figure 31.1

The Assembled Device

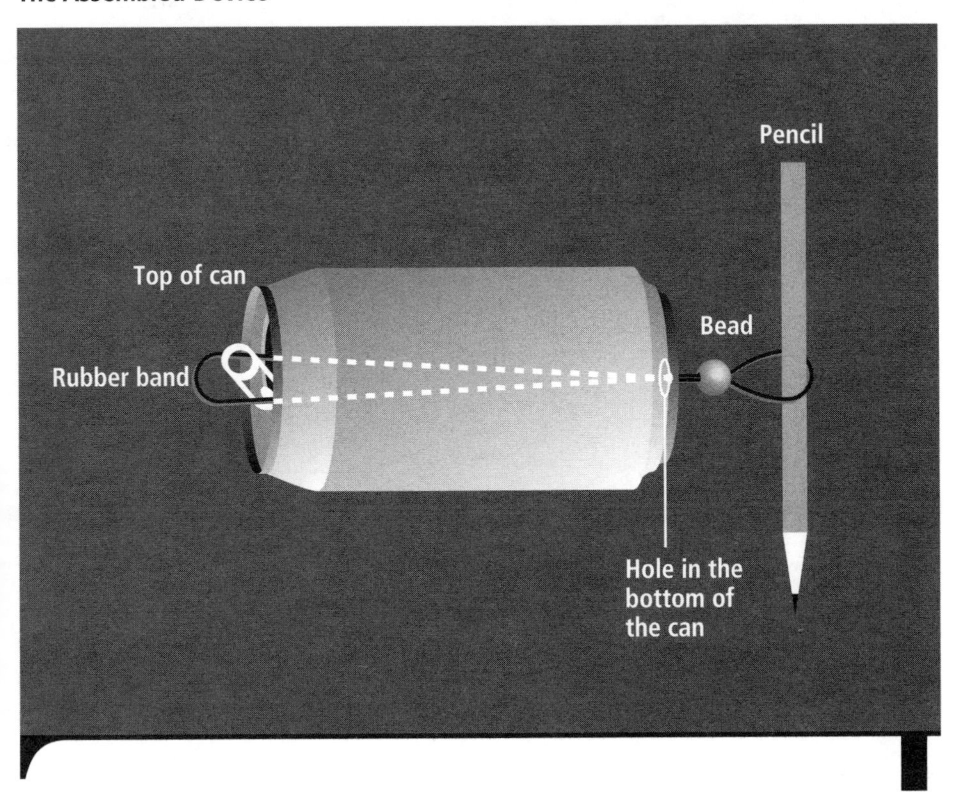

4. Now spin the pencil 50 to 100 times and put the can on the floor and observe.
5. If the can spins out, you can put a rubber band around the can opposite the end with the pencil to give it more traction. If the pencil spins out, you can use a longer pencil or slide the pencil so that most of it is on one side.
6. Find ways to make the can go faster and farther. You can try things such as using different rubber bands, reducing friction, and using a spool of thread instead of a soda can.

Post-Lab Questions

1. What modifications did you have on your final racer, and how did they help?
2. Explain how the car works. Use the terms *potential energy, kinetic energy, elastic,* and *conversion* in your explanation.
3. If the car rolls forward once for each turn of the pencil, how far should the car go if you turn it 50 times?

Extensions

Build other types of vehicles that convert one type of energy to motion. Use mouse traps, springs, weights, or other objects to make them move. See how many different designs you can think of that use different kinds of energy.

LAB 32: BERNOULLI'S PRINCIPLE

In this lab, students will use a little background information about Bernoulli's principle to figure out how the spinning of a moving ball affects its trajectory. The activity is inquiry in that students will be discovering this relationship on their own. Those with baseball, tennis, or soccer experience may already have some intuition as to how this works.

It takes some coordination to get the ball to roll down the ruler and then throw it with a whipping motion. Practice will help. If a student cannot get this to work, he or she can use the cardboard tube at the center of a roll of paper towels to make it easier.

SCiLINKS THE WORLD'S A CLICK AWAY
Topic: Bernoulli's Principle
Go to: *www.sciLINKS.org*
Code: THP20

Post-Lab Answers

1. Step 2: up; Step 3: right (left for lefties); Step 4: left (right for lefties); Step 5 (very difficult to accomplish): down
2. The ball will deflect left if spinning counter-clockwise (from above) and right if clockwise. It will curve up with backspin and down with topspin. General rule: The ball will curve toward the side of the ball that is spinning back toward the person throwing the ball.
3. Diagrams should show the side spinning into the "wind" having a higher relative speed and lower pressure than the side spinning away from the "wind."

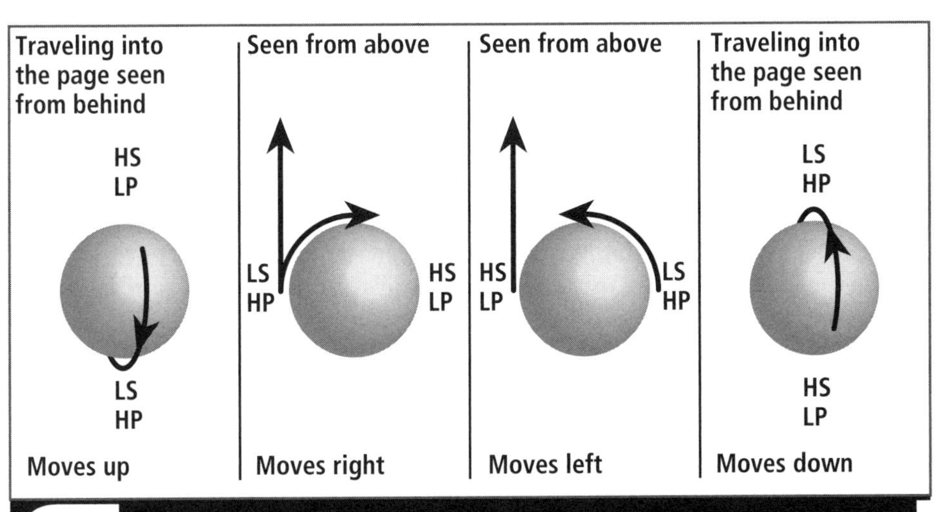

HS=High Speed LS=Low Speed HP=High Pressure LP=Low Pressure

LAB 32: BERNOULLI'S PRINCIPLE

QUESTION

How does spin affect the motion of a projectile?

SAFETY

It is best to do this activity outside on a day with no wind to avoid hitting someone or breaking something.

MATERIALS

Plastic ruler with groove down the middle or a cardboard paper-towel tube, Ping-Pong ball

PROCEDURE

Air exerts pressure on surfaces. Bernoulli's principle says that when air is moving parallel to a surface, it does not impart much force to that surface. The faster the air is moving, the less force it imparts. This is part of the explanation for how an airplane wing works (it also has to do with "angle of attack"). It's also why a "ragtop" convertible top on a car bulges upward on the freeway. When the car is at rest, air pressure inside and outside are equal, and the top stays flat. As the car speeds up, the fast flowing air along the top of the car exerts less pressure than when at rest, while the air pressure inside the car remains constant. There is more pressure upward than downward, so the soft top bulges upward.

When an object is spinning and moving (see Figure 32.1), the air moves across one side faster than on the other side of the spinning object, so a similar force is created. In this activity, you will determine how the direction of spin is related to the direction that the ball curves. Keep in mind that when a ball moves through the air, there is "wind" flowing past it. If the ball is spinning into that "wind," it slows it down. If it is spinning away from that "wind," it speeds it up.

Note that there is another explanation of how a ball curves that relies on the Magnus effect. It says that the spinning ball throws air in one direction and action/reaction moves the ball in the opposite direction. It is likely that both effects contribute to the motion.

Remember that Bernoulli's principle only applies when the air is flowing along a surface, not when it collides directly with it. The front grill on a car does not experience Bernoulli's forces because the air collides with it directly.

Figure 32.1

Diagram of a Spinning Ball

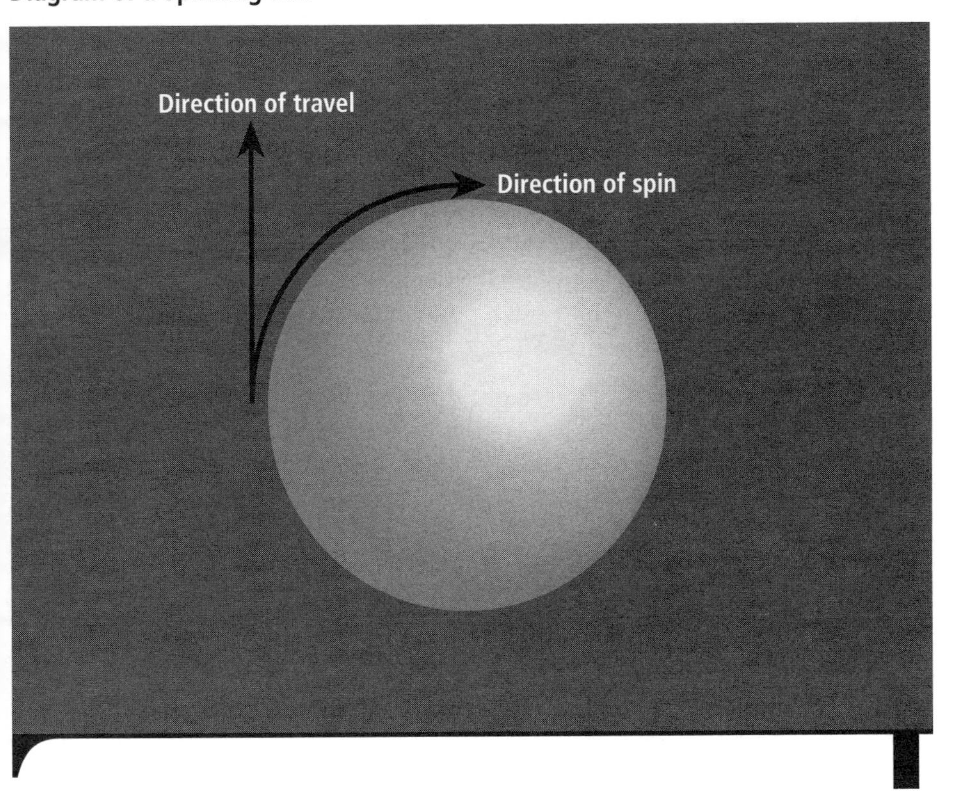

1. You will be using a ruler with a groove in it to throw a Ping-Pong ball. Allow the ball to roll quickly along the ruler and launch it with a "whipping" motion to create a high spin.
2. Throw the ball overhand so that the bottom is spinning in the direction of motion (backspin). Record whether the ball moves up, down, left, or right compared to a spinless throw.
3. Throw the ball sidearm so that the side closest to you is spinning in the direction of motion. Record whether the ball moves up, down, left, or right compared to a spinless throw.
4. Try to throw the ball with the opposite spin from Step 3. Either use the same hand but with a backhand motion (your throwing arm starts across

your body) or use the other hand. Record whether the ball moves up, down, left, or right compared to a spinless throw.

5. Now attempt to throw the ball underhand so that the top of the ball is spinning in the direction of motion. Record whether the ball moves up, down, left, or right compared to a spinless throw.
6. If you have difficulty doing this with a ruler, use a cardboard paper towel tube.

Post-Lab Questions

1. Looking at the ball from above, draw an arrow on each one of the diagrams below to show which way the ball was diverted from its normal path.

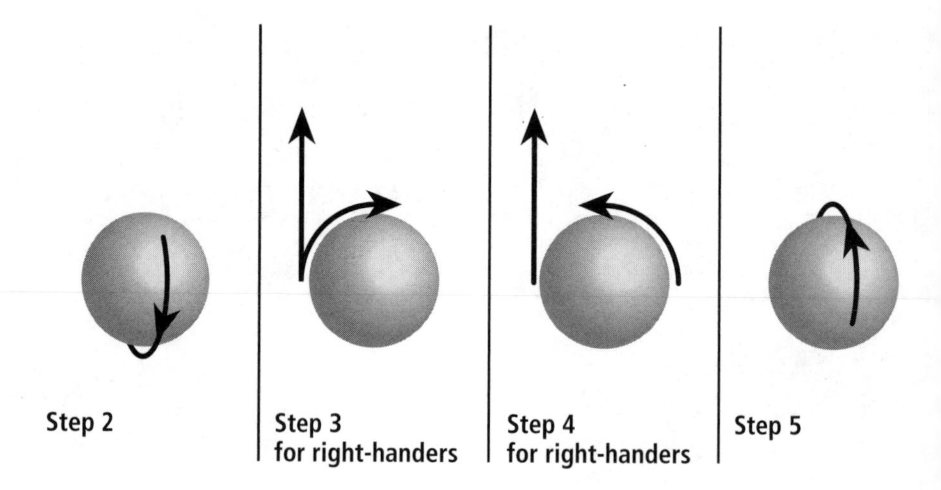

2. Write a general rule for which direction the ball will move compared to its spin.
3. Draw a diagram for Steps 2, 3, 4, and 5 similar to those in question 1 and include:
   a. direction of spin
   b. observed direction of motion
   c. side of the ball where the air is moving faster relative to the spinning ball (top or bottom for Steps 2 and 5 and left or right for Steps 3 and 4)
   d. side of the ball where the air is moving slower relative to the spinning ball
   e. direction that Bernoulli's principle predicts the ball will move based on the difference in pressure

   Did the direction that Bernoulli's principle predicts match your observations in question 1?

LAB 33: BUOYANCY 1

In this lab, students will see that for any object that sinks, the object displaces more water when it is floating than when it is submerged. This lab is inquiry in that students do not know the answer to the question before they begin. The teacher could add more inquiry if desired by allowing the student to design the procedure independently.

Topic: Buoyancy
Go to: *www.sciLINKS.org*
Code: THP21

Post-Lab Answers

1. If the rock sinks, its density is greater than 1. This means that its mass is higher than its volume.
2. It should displace more water when floating because that's where it displaces its mass instead of its volume.
3. An object displaces an amount of water equal to its mass when it is floating and a volume of water equal to its own volume when submerged underwater. Because a rock has a density greater than 1, its mass is greater than its volume. Because it displaces its own mass when floating, it will displace more water when it is floating.

LAB 33: BUOYANCY 1

QUESTION

A boat with a large rock in it is floating in a lake. If the rock is thrown into the water, will the level of the lake go up, go down, or stay the same?

SAFETY

Standard safety precautions apply.

MATERIALS

Small plastic salsa cup, transparent drinking glass, rock, nonpermanent marking pen

PROCEDURE

To understand the question in this lab, one must know about buoyancy and density. There are two important principles of buoyancy:

1. When an object is floating, it displaces a volume of water equal to its own mass.
2. When an object is underwater, it displaces a volume of water equal to its own volume.

   Hypothesis: The water level will ________________ because _________

   ____

For this activity, you will determine your own procedure for answering the question, determine what data should be collected, and draw your own conclusion. Write a paragraph describing your procedure, draw at least one diagram showing your setup, and create a data chart to record your data.

Post-Lab Questions

1. Use what you know about density to explain your answer to this question: Which number is larger, the volume of a rock in cm^3 or the mass of the same rock in grams?
2. Based on your answer to #1, in which position should the rock displace more water, floating or underwater?
3. Write a paragraph that completely explains why the results of this experiment come out the way that they do.

Extension

If a 2 m^3 rock with a density of 3,000 kg/m^3 is dropped into a lake from a boat, how much will the volume of the lake change if the density of the water is 1,000 kg/m^3?

LAB 34: BUOYANCY 2

This activity is inquiry in that the students do not know the answer to the question before they begin. Students will use the calibrated rubber band that they created in Lab 27: Spring Constants. They will see that an object is lighter when submerged, and the more dense the liquid, the lighter the object will be. This is important in understanding how a huge ship can float in water, how a hot air balloon floats in air, and how a submarine works.

Note that you need either heavy weights or thin rubber bands to distinguish between water and oil.

Post-Lab Answers

1. The washers appear lighter in water than they are in air.
2. The washers appear lighter in water than they are in oil.
3. Because oil floats on water, water is more dense than oil. The more dense the fluid is, the more weight the object will appear to lose when submerged. Because water is more dense, the washers are lighter in water than they are in oil or in air.
4. Because mercury is more dense than the washers, they will float in the mercury and the scale will not read any weight. The scale will read zero.

LAB 34: BUOYANCY 2

QUESTION

How does the apparent weight of an object change when it is submerged?

SAFETY

Do not drink the water or use the oil for cooking when finished.

MATERIALS

2 small salsa cups, cooking oil, water, 4 large washers, calibrated rubber band (created in Lab 27)

PROCEDURE

The term *buoyancy* is used to describe the force that is applied on an object submerged in a fluid. A helium balloon floats because the buoyant force of the air pushing up on the balloon is greater than the gravitational force pulling down on it. A beach ball floats on water because the ball is lighter than an equal volume of water. For example, the beach ball might have a volume of 5 l and a mass of 50 g; 5 l of water would have a mass of 5,000 g. Because the ball weighs less per unit volume (has lower density) than the water, it floats.

1. Pour water into a cup to a depth of about 4 cm. Pour cooking oil into another cup to a depth of about 4 cm.
2. Hang four large washers from a paper clip attached to your calibrated rubber band. Measure the calibration mark carefully and record the results here. ___________ cm

3. Submerge the washers in the water, measure the calibration mark carefully, and record the results here. Be sure that the washers don't touch the bottom. __________ cm
4. Submerge the washers in the oil, measure the calibration mark carefully, and record the results here. Don't allow the washers to touch the bottom. __________ cm
5. You need to know which is more dense, oil or water. Mix them and see.
6. Clean the washers with dishwashing liquid to remove the oil, and then dry them completely before putting them away.

Post-Lab Questions

1. How did the weight of the washers in air compare with the apparent weight in the water?
2. How did the apparent weight of the washers in the water compare with the apparent weight in oil?
3. Explain your answers to numbers 1 and 2 using what you know about density and buoyancy.
4. What would the reading on the calibrated rubber band be if you tried this with mercury, which has a higher density than the washers? (*Warning*: *Do not try this experiment. Mercury is poisonous.*)

Extension

Find the buoyant force on a helium balloon by attaching pieces of duct tape to the balloon until it becomes neutrally buoyant (doesn't float, doesn't sink). Now pop the balloon to let the helium escape, and then find the balloon's weight with the duct tape. When an object is neutrally buoyant, the buoyant force is equal to the weight of the object. The helium that escaped when you popped the balloon has some mass, but it is small and can be ignored (approximately 0.30 g depending on the size of the balloon).

LAB 35: BUOYANCY 3

This activity is inquiry in that students do not know the answer before they begin. If you were to poll the class before doing this activity, you would probably find that the hypotheses are split down the middle of the class.

It is important that small, thin cups are used so that the mass of the washers is on the same magnitude as the cup of water. If the water is not very deep in the cups, you might need to use more than three washers to get more metal under water.

See this website for a video of a similar experiment: *http://groups.physics.umn.edu/demo/fluids/movies/2B4015.mov*. If the website moves, do a search online for "finger water balance" and find a video of a person sticking his or her finger in a beaker of water on a balance.

Post-Lab Answers

1. Yes, the weight of the cup of water changes slightly, but not as much as the washers themselves. It changes less than that, possibly by the weight of the water displaced.
2. For an object that is floating or an object resting on the bottom, the weight will change by the weight of the object.
3. Yes, the rock will get somewhat lighter underwater.

LAB 35: BUOYANCY 3

QUESTION

Does the mass of a container of liquid change when an object is suspended in it?

SAFETY

Do not use any of the liquids when finished.

MATERIALS

2 small plastic salsa cups, ruler, pencil, water, small spool of thread, 3 washers

PROCEDURE

In this experiment, you will see how the weight of the container of fluid changes when an object is submerged in it. It's obvious that if you drop washers down to the bottom of a cup of water, the cup of water's weight will increase by the weight of the washers. How about if you dangle the washers in the water without touching the bottom? Does an aquarium get heavier when you put swimming fish in it?

Hypothesis: I think that the weight of the cup of water will ______________ ____________________.

1. Set up your ruler with two small cups of water on each end so that it is in balance. The fulcrum can be a regular six-sided pencil. The flat sides will help you balance the ruler, but will not affect the results too much.
2. Hang three washers from a thread (~10 cm) and submerge them in the water on one side of the balance without touching the bottom. Record your observations.
3. Have a partner hold a small cup of water in her hand and close her eyes. Dip the washers in the water without touching the bottom and see if

she can feel when the washers go into the water. Dip them in at random intervals several times and see if your partner can detect it. Repeat the procedure, this time letting the washers touch the bottom.

Post-Lab Questions

1. Does the weight of the cup of water change when the washers are submerged? Does it appear to change by the entire weight of the washers? Or does it appear to change less than that?
2. How do you think the weight of a lake changes when a large boat floats in it? How does it change when a rock is dropped to the bottom?
3. If a rock that was just a little too heavy to lift when not in the water was put into the water, would you be able to lift it?

Extension

Find out if the results of this experiment would have been different if the string had been attached to the lid of the cup. Make sure that the washers are hanging from the lid, the lid is on the cup, and that the washers are not touching the bottom. Perform the experiment and see for yourself.

LAB 36: HERO'S ENGINE

In learning how this apparatus works, students will improve their understanding of action/reaction. This knowledge will also help students understand how a rocket engine works and many other important phenomena. Just as the exhaust coming from the rear of a rocket propels the rocket forward, the water coming from the Hero's Engine tangentially causes the can to spin in the opposite direction.

Post-Lab Answers

1. The can rotates the opposite direction of the water coming out. If the water comes out in a clockwise direction, the can rotates counter-clockwise.
2. Newton's third law, which states that for every action, there is an equal and opposite reaction.
3. When the water comes out of the hole, say, in a counter-clockwise direction, it must apply a force to the can in a clockwise direction. The action of the water coming out creates a reaction on the can that applies a torque to the can and causes it to spin.

LAB 36: HERO'S ENGINE

QUESTION

How does a Hero's Engine work?

SAFETY

Be very careful when puncturing the can with a nail. Wear gloves or get an adult's help if you feel that you cannot do this safely.

MATERIALS

Empty soda can, thread, iron nail

PROCEDURE

A Hero's Engine is a device that propels a fluid at an angle through holes from a round object hanging freely. The object will start to spin when the fluid comes out (see Figure 36.1). This is similar to how a rocket engine works except that the exhaust propels the rocket instead of spinning it.

1. Using a sharp nail, poke a hole in the bottom of an empty soda can and then press the end of the nail such that the hole is pointing off to the side. If this does not work well, glue short pieces of straws in the holes, pointing at an angle.
2. Do the same thing on the opposite side of the can, making sure that the two holes are pointing in the same direction along the circle.
3. Tie a 10 cm thread to the tab on top of the can and try your best to bend the tab so that the thread is coming straight up from the center of the can.
4. Fill the can with water and cover the holes with your fingers.
5. Hold the can by the thread and remove your fingers, allowing the water to pour out over the sink.

Figure 36.1

Hero's Engine Model

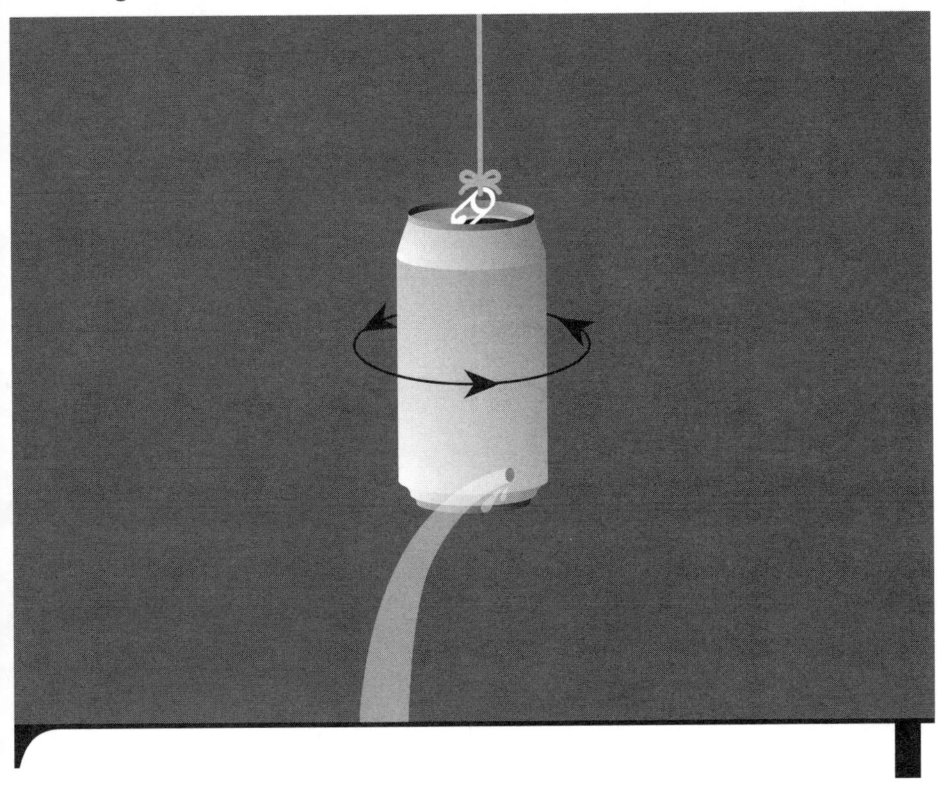

Post-Lab Questions

1. Did the can rotate in the same direction that the water was coming out or the opposite?
2. Which of Newton's laws is responsible for this action?
3. Explain in your own words how this demonstration works. Use the terms *force, action, reaction*, and *torque* if possible.

Extension

Using the same can as in the lab but with no water in it, weigh it down with sand or pennies so that it will sink when placed in water. Prepare a bucket of water, hold the weighted can by the string, and submerge it in water. Which way does the can spin? Why?

LAB 37: PRESSURE

This lab is not inquiry-based because the teacher must explain how to calculate pressure before doing the activity, but the lab gives students a great concept of what PSI means. They usually have no idea if 10 PSI is a lot or a little. They also don't understand how a small pressure over a large area can create a large force. This lab helps them understand those concepts. There are great demonstrations to show the incredible lifting power of small pressures over large areas. Blowing up a trash bag with straws to lift a student is an example.

Topic: Pressure
Go to: *www.sciLINKS.org*
Code: THP22

Post-Lab Answers

1. Answers will vary, but are normally in the 10–20% range, mostly because of items in the trunk of the car.
2. Answers will vary. A 200-pound person with one foot that covers 40 square inches will exert a pressure of 2.5 PSI.
3. Answers will vary. Based on the 200-pound person above, the pressure would be 5 PSI.

LAB 37: PRESSURE

QUESTION

What is the pressure exerted by your weight on the bottom of your feet? How can pressure on a car's tires be used to determine the car's weight?

SAFETY

Be extremely careful working around a car. Be sure that it is on flat ground, the parking brake is on, and the gear is in park.

MATERIALS

Graph paper, chalk, ruler, bathroom scale

PROCEDURE

Pressure depends on the force being applied, as well as the area over which it is spread. A large force applied to a small area creates a very large pressure. If you hit a hammer on a nickel, the nickel will not pierce a piece of wood because the pressure is not high enough. If you hit a hammer on a nail, it will pierce a piece of wood because the tip of the nail is so small. Pressure can be reported in many different units, including Pascals, millimeters of mercury (mmHg), pounds per square inch (PSI), torr, and atmospheres (atm). A weather reporter gives the barometric pressure in inches of mercury. A car tire's pressure is measured in PSI. Because PSI is square inches, we will do this activity without using metric measurements.

In this lab, you will calculate the pressure on the bottom of your foot and use tire pressure to calculate the weight of a car.

Part 1

1. Weigh yourself on the bathroom scale and record it in a data chart similar to the one on page 153.

2. Trace your shoe on a piece of graph paper and count how many squares your shoe covers. Use the number of squares and the area of each square to calculate the area of your foot. If the graph paper is 4 squares per inch, then each square is 1/16 of a square inch.
3. Multiply that number by 2 to include your other foot.
4. Divide your weight in pounds by the area of your feet to calculate the pressure on the bottom of your feet in PSI.

Part 2

1. Measure the pressure in each of the four tires on a car with a pressure gauge, or use the recommended tire pressure as an estimate.
2. Check that the parking brake on the car is set and the car is in park.
3. Measure the outline of the part of the tire that touches the ground and calculate the area or the rectangle that it makes.
4. Multiply each area times the pressure in that tire to figure out the weight held up by that tire.
5. Add the weights supported by each tire to find the weight of the car.

Data Chart

Part 1

Student's weight ______________ pounds

# of graph paper squares for one foot ______________ squares

# of square inches for one foot ______________ in^2

# of square inches for two feet ______________ in^2

Pressure on the bottom of two feet ______________ PSI

Part 2

Front left tire: pressure _____ PSI, area of contact _____ in^2, weight supported _____ pounds

Front right tire: pressure _____ PSI, area of contact _____ in^2, weight supported _____ pounds

Rear left tire: pressure _____ PSI, area of contact _____ in^2, weight supported _____ pounds

Rear right tire: pressure _____ PSI, area of contact _____ in^2, weight supported _____ pounds

Post-Lab Questions

1. Check the owner's manual or search the internet and try to find the correct weight of the car. Search for something like "curb weight Honda Civic 2008." What was your percentage error?
2. Atmospheric pressure (the weight of the air spread out over the ground) is approximately 15 PSI. How does the pressure on the bottom of your feet compare to that?
3. What would be the pressure on the bottom of your shoe if you were wearing high heels with half the area of a flat shoe?

Extensions

It is said that egg shells are strong enough that if you could cut them in half and lay them on the ground, you could walk across them without breaking them. Try it. If you don't want to use real eggs, use plastic eggs from the holiday section of the store.

How do people lay on beds of nails without being injured? Does it take mystical powers? Try this with a balloon. Put 100 tacks through a piece of cardboard about ½ inch apart and then press a balloon down on it.

LAB 38: PRESSURE VERSUS DEPTH

In this inquiry activity, students do not know the equation for pressure versus depth before beginning, but through guided activities they discover it themselves. This is a powerful way to get students to conceptually understand the equation and remember it for a long period of time. But the first time that you do an activity like this, you may be disappointed in the results. The students have probably never done an activity like this, and when test time comes, they likely won't reflect on the activity to be able to solve these problems. The first couple of times that you use discovery/inquiry learning, you will have to remind students of the activity. A test question may read, "Recalling the pressure versus depth lab, what is..." Eventually you can stop using that stem as the students become used to the idea of learning from a lab and reflecting on the lab during assessments. The state assessments will not have those stems, so you do have to wean students away from answering questions with stems. So don't be shocked when performance is not as you expected the first couple of times. It is not a learning issue; it is a disconnect between hands-on learning and assessments.

If the teacher anticipates that students will not be able to connect increased pressure with an increase in how far the liquid travels, a prelab may be used. Have students blow gently in a water-filled straw and then blow hard into the straw and see how far the water travels.

Topic: Fluids and Pressure
Go to: *www.sciLINKS.org*
Code: THP23

Answers to Procedure Questions

Part 1

Students should notice that the water shoots farthest from the bottom hole.

Observation 1

The pressure of a fluid increases as its depth increases. This means that pressure and depth (h) are directly related, which can be represented by $P \alpha h$.

Part 2

1. Cold air will flow out of the refrigerator, which is why it is bad to hold the door open. It does not flow very quickly.
2. If the refrigerator were filled with water, the water would flow out much faster than air would.

3. It would not shoot out as far.
4. It would shoot out farther.
5. It would not shoot out at all because there is no pressure difference between inside and outside.
6. Water is more dense than air.

Conclusion 2

The pressure of a fluid increases as its density increases. This means that pressure and density (ρ) are directly related, which can be represented by $P \alpha \rho$, where ρ represents density.

Part 3

1. The water stops coming out because the simulated zero-g makes it so that gravity is not pulling out the liquid. *Note:* What is really happening is that the can is accelerating just as fast as the water, so the water does not fall relatively faster than the can. Again, this is simulated zero-g, not actual zero-g. This is the same reason that astronauts float around in the space shuttle: The space shuttle is falling and the astronauts are falling so they do not fall relative to each other. It is not because there is no gravity in space.
2. On the moon, where gravity is weaker, the water would not shoot out as far.
3. On Jupiter, the water would shoot out farther (assuming that it didn't freeze).

Conclusion 3

The pressure of a fluid increases as gravity increases. This means that pressure and gravity are directly related, which can be represented by $P \alpha g$.

Post-Lab Answers

1. Pressure = (density)(gravity)(depth)
2. Pressure = $(1{,}000 \text{ kg/m}^3)(9.8 \text{ m/s}^2)(40 \text{ m})$ = 392,000 Pascals + 101,300 Pascals = 493,300 Pascals
3. Pressure = $(1{,}150 \text{ kg/m}^3)(9.8 \text{ m/s}^2)(10 \text{ m})$ = 112,700 Pascals + 101,300 Pascals = 214,000 Pascals

LAB 38: PRESSURE VERSUS DEPTH

QUESTION

How does the depth of a liquid affect the pressure that it exerts?

SAFETY

Be extremely careful punching holes in the can. The nail and the can will be sharp.

MATERIALS

Empty aluminum soda can, nail

PROCEDURE

In this lab, you will be figuring out the equation for how pressure changes with depth in a fluid using a combination of observation and deduction. Remember that a fluid and a liquid are not the same thing. A fluid is something that flows, and that includes liquids and gases.

Part 1

1. Find an aluminum can and use a sharp object to carefully poke three holes in the can in a vertical line (see Figure 38.1, p. 158).
2. Predict what will happen if you fill the can with water and let the water flow out of the holes. Which hole will flow the farthest, or will they all flow the same distance?
3. Tip the can back and fill it with water above the top hole.
4. Hold the can over the sink and allow the water to flow out.
5. Sketch what you observed viewing the can from the side.

Figure 38.1

Can With Three Holes

Observation 1

The pressure of a fluid ________ (increases, decreases, stays the same) as its depth increases. This means that pressure and depth (h) are _______ (directly, inversely) related, which can be represented by _________ ($P \alpha h$, $P \alpha 1/h$).

Part 2

1. Take off your shoes and socks and go into the kitchen. Open the refrigerator door. What do you feel?
2. Imagine the refrigerator was filled with water, which is much more dense than air. Would the water pour out faster than the cold air?
3. Imagine that your can was filled with alcohol, which is less dense than water. Based on pressure alone, would the alcohol shoot out farther than or not as far as the water?
4. Imagine that your can was filled with mercury, which is much more dense than water. Based on pressure alone, would the mercury shoot out farther than or not as far as the water?
5. Imagine that the can was filled with air, which is much less dense than water. What would happen?

6. When you go from the top of a tall building to the bottom (hundreds of feet), your ears do not hurt. When you go 5 ft. underwater, your ears hurt from the pressure. Explain why.

Conclusion 2

The pressure of a fluid __________ (increases, decreases, stays the same) as its density increases. This means that pressure and density (ρ) are ________ (directly, inversely) related, which can be represented by ____________ ($P \alpha \rho$, $P \alpha 1/\rho$), where ρ represents density.

Part 3

1. An object can experience apparent zero gravity when it is falling, similar to astronauts in orbit. Go outside and fill your can with water and cover the holes. Hold the can out and just as you uncover the holes, drop it. What happens?
2. Imagine that you took your can to the Moon, where gravity is much weaker. The water would shoot out ________ (farther than, less far than, the same as) on Earth?
3. Imagine that you took your can to Jupiter, where gravity is much stronger. Would the water shoot out ________ (farther than, less far than, the same as) on Earth?

Conclusion 3

The pressure of a fluid __________ (increases, decreases, stays the same) as gravity increases. This means that pressure and gravity are ________ (directly, inversely) related, which can be represented by ______________ ($P \alpha g$, $P \alpha 1/g$).

Post-Lab Questions

1. To write the equation for pressure, put all the variables that were direct relationships on the top of the fraction and all the variables that were inverse relationships on the bottom.

   Pressure = ____________________

2. Using your new equation, calculate what the pressure would be at a depth of 40 m in water (density = 1,000 kg/m^3). Your answer will come out in Pascals, and don't forget that the pressure at the surface of the water is 101,300 Pa from atmospheric pressure so you must add that to the pressure of the water.
3. Now calculate what the pressure would be at 10 m below the surface of saltwater in the ocean (density = 1,150 kg/m^3).

Extensions

Every 10 m underwater that you go, the pressure increases 1 atm. What is the deepest under the ocean that a human has ever gone in a submersible vehicle? What is the pressure down there? What is the deepest that a diver has gone without any breathing apparatus? What is the pressure down there? How deep can a penguin dive? What is the pressure down there? What is the cruising altitude for passenger airplanes? Do research to find what the atmospheric pressure is up there.

LAB 39: THERMODYNAMICS

This is an inquiry activity in that there is no one correct answer. Each student will come up with a different solution with different levels of success. Students should have been exposed to conduction, convection, and radiation in elementary or middle school and be able to use that information to minimize heat loss through the use of insulation.

Make the water for the competition as hot as possible without being dangerous. This will give you better resolution to determine the winner. Because every thermometer is a little different, be sure to write down the beginning temperature and the starting time. It will be difficult to start 36 students at exactly the same moment, so just write down the time and read each thermometer at 20 minutes.

Topic: Thermodynamics
Go to: *www.sciLINKS.org*
Code: THP24

Post-Lab Answers

1. It cannot heat through conduction because we are not touching. It cannot heat through convection because there is no air between us.
2. An insulator does not affect radiation much. The can still radiates heat into the insulation, but at least the heat doesn't escape completely. The insulation keeps the air from touching the can, which reduces both conduction and convection.
3. Polar bears have transparent hair and dark skin. That lets the radiation hit the skin and warm it, but the hair insulates the body to reduce conduction and convection. This is a very efficient way to increase the input of energy and decrease the output. Some sources say that the hair of the polar bear is like fiber optics and guides the light to the skin. This is not correct. It is simply transparent. Other animals have layers of blubber to insulate the important organs in the body.

LAB 39: THERMODYNAMICS

QUESTION

What is the best insulator to keep water warm the longest?

SAFETY

Do not drink the water after the experiments.

MATERIALS

Empty soda can, microwave or stove to heat water, variety of materials to use as insulation

PROCEDURE

Heat can be transferred by three means: conduction, radiation, and convection. *Conduction* is heat transfer via direct touching. If you touch a hot stove, heat is transferred from the stove to your hand by conduction. *Radiation* is heating without any contact at all. If you stand several feet away from a bonfire, you can still feel the heat through radiation. This is not the same as nuclear radiation. The sun heats a person via radiation. *Convection* is heat transfer via heating surrounding fluid (gas or liquid) and then the fluid moves. It helps explain how a hot oven can heat up the entire kitchen. Heat radiates from the front of the oven, heats the air, and then the air swirls around the kitchen and heats the entire room.

In this lab, you will be practicing for a contest that will be held in class. Following are the rules for the contest:

1. You will be insulating an aluminum soda can such that hot water put into your can will stay hot longer than everyone else's water.
2. Your design should be based on both research and experimentation.
3. Your design should have a standard aluminum can with the opening exposed so that the water can be poured in and a thermometer can be inserted.

4. Your apparatus must be no larger than 20 cm × 20 cm × 20 cm.
5. Your apparatus cannot use any heat source of its own. It can only insulate the heat of the water put into it.
6. Hot water and a thermometer will be placed into your apparatus. Twenty minutes later, the temperature will be compared to the others in the class and the highest temperature wins.
7. You have one week to research and experiment with insulation before the contest begins. Your grade will be based on how little the temperature of your water changes in the 20 minutes.
8. Record the results of each one of your trials any way that you deem most appropriate.

Post-Lab Questions

1. How can we be so sure that the sun does not heat us through convection or conduction?
2. Which of the three means of heat transfer do you think is most important to minimize in this contest in order to win? Explain why.
3. How do mammals that live in cold climates and cold water insulate themselves?

Extension

There is a common physics question about transfer of heat. You order a cup of coffee and just as it shows up, you realize that you need to use the restroom. In order to keep the coffee as hot as possible while you're gone, should you add the cream to the coffee now or when you get back? Explain why. Set up a controlled experiment and try this.

SECTION 3:

Waves, Sound, and Light

LAB 40: CENTER OF PERCUSSION

This activity is inquiry in that students cannot locate or define *center of percussion* before the lab. Even though it appears to students that the point of this lab is to find the center of percussion, the real point is to learn about the period of a pendulum and see how length affects the period of that pendulum. Doing this through a real-world application just adds engagement to the activity.

Post-Lab Answers

1. ($^2/_3$) = 0.67 or 67% of the length. The ruler should act similarly to a physical pendulum. Because there is some ruler above the rotation point, it will not be perfect.
2. Not really. The center of percussion is a point, not an area. Whatever is causing the bat to seem like it has a bigger sweet spot is not related to center of percussion. It might have some vibration dampening in it to keep it from vibrating if you don't hit the ball with the sweet spot.
3. It would be 60 cm from the handle. When a person swings the bat, their hands may affect the location of the sweet spot. Also, the bat is not a rod: Its mass is concentrated toward its tip. Both of these effects should move the sweet spot closer to the end of the bat (farther from the hands).

LAB 40: CENTER OF PERCUSSION

QUESTION

How can you determine the center of percussion or "sweet spot" of a baseball bat?

SAFETY

Standard safety precautions apply.

MATERIALS

Ruler, thread, washers

PROCEDURE

If you've ever played baseball or softball, you know that there is a place near the end of the bat that when you hit the ball there, it does not hurt your hands as much as when you hit it at other locations. In everyday terms, this is called the *sweet spot*. A tennis racket has a sweet spot as well. There are many different definitions of *sweet spot*. Here, we will use the spot where hitting the ball causes the least discomfort for the batter. The physics term for this location is the *center of percussion*. The center of percussion is the spot where if you strike the object at this point, there will be no vibrations at the pivot point.

You can locate the center of percussion of a physical pendulum (a solid pendulum) by comparing its period with that of a standard pendulum (weight hanging from a string). To do so, you just find the length of the standard pendulum that will give you the same period as the physical pendulum. Once the two pendulums swing in unison, put a mark on the ruler where the center of the mass on the standard pendulum aligns with the ruler, and this is the sweet spot.

1. Put a thin rod through the hole in the end of your plastic ruler and have an assistant hold it while it is swinging. Measure the length of the physical pendulum from the top of the hole to the bottom of the ruler. __________ cm
2. Tie three washers to the end of a thin thread at least as long as the ruler. Hold it next to the physical pendulum and set them both in motion.
3. Adjust the length of the string until the two pendulums have the same period. Let them swing at least five swings in unison. Once you accomplish this, measure the length of the string from where you're holding it to the center of the washers (their center of mass). __________ cm

Post-Lab Questions

1. What is the ratio of the length of the standard pendulum to the length of the physical pendulum? If the physical pendulum were a rod, the ratio would be 2/3 (two-thirds). Does your ruler act like a rod?
2. Some baseball bats advertise, "Bigger sweet spot than our competitors." Based on the definition of the sweet spot of a baseball bat being located at the center of percussion, does this make sense?
3. If your ratio in #1 works for a baseball bat, too, where would the sweet spot be on a 90 cm bat?

Extension

Do an experiment to find the spot on the baseball bat that creates the weakest vibrations when struck there (another definition of sweet spot). Does the result match your model (two-thirds the length of the bat)?

LAB 41: SOUND WAVES

In this inquiry activity, students will discover the relationship between length and pitch (frequency), tension and pitch, and mass per unit length and pitch without being instructed first. Many students confuse pitch and volume, so the teacher may have to demonstrate the difference in class before assigning this activity.

Topic: Interactions of Sound Waves
Go to: *www.sciLINKS.org*
Code: THP25

Post-Lab Answers

1. Longer lengths produce lower-pitched sounds (frequencies). Shorter lengths produce higher-pitched sounds.
2. Higher tension produces higher-pitched sounds. Lower tensions produce lower-pitch sounds.
3. Higher mass (per unit length) produces lower frequencies. Lower mass per unit length produces higher frequencies.

LAB 41: SOUND WAVES

QUESTION

How do tension, length, and mass per unit length affect the frequency of a vibration?

SAFETY

Standard safety precautions apply.

MATERIALS

Thick rubber band, thin rubber band, glass or plastic soda bottle, ruler

PROCEDURE

Vibrating objects that compress and stretch masses of air create sound waves. Your vocal cords vibrate, the head of a drum vibrates, a clarinet reed vibrates, and the resonator of a guitar vibrates to make sound.

The pitch (tone, musical note) that comes out of a wind instrument, a percussion instrument, and a string instrument are all controlled by different factors.

1. Place the plastic ruler over the edge of the table so that 3 inches are hanging off the table. Pluck the ruler and notice the note that it emits.
2. Allow 3 inches to hang over and pluck it again. Did the pitch get higher or lower?
3. Allow 6 inches to hang over and pluck it again. Did the pitch get higher or lower?
4. Hold the thin rubber band between a finger and your thumb and stretch it lightly. Pluck it and notice the note it emits.
5. Stretch the rubber band tighter and pluck it again. Did the pitch get higher or lower?

6. Stretch it even more and pluck it again. Did the pitch get higher or lower?
7. Perform the same experiment with the thick rubber band. Were the notes higher or lower pitched than when you stretched the thin rubber band by the same amount? You can put them both side by side on your fingers and compare.
8. Practice blowing across the top of your bottle to produce a musical note. The bottle should hang straight down and touch your lip. You should blow straight across the top of the bottle smoothly and gently. Notice the note that it produces.
9. Now fill the bottle ⅓ with water and blow across it again. Did the pitch get higher or lower?
10. Now fill the bottle ½ with water and blow across it again. Did the pitch get higher or lower?
11. Now fill the bottle ⅔ with water and blow across it again. Did the pitch get higher or lower?
12. Arrange your data in a manner that you deem appropriate.

Post-Lab Questions

1. How does the length of the vibrating object affect its frequency?
2. How does the tension of a vibrating object affect its frequency?
3. How does the mass of a vibrating object affect its frequency?

Extension

Find someone who has a guitar or other string instrument and see how your analysis stands up. See which string is thickest: the high frequency (high pitch; treble) or the low frequency (low pitch; bass). With the person's permission, change the tension in a string and see how it affects the pitch. When you push on one of the strings (shortening the string), how does the frequency change?

LAB 42: REFRACTION OF SOUND

This lab introduces students to refraction using something they can experience. They really do not understand what is going on when a pencil is put into water and appears to bend. It is difficult to see rays of light bending without the use of a laser or other expensive equipment. So for this first exposure to refraction, students will use sound instead.

Post-Lab Answers

1. When sound travels from air through the CO_2-filled balloon, it gets focused. When sound travels from air through the helium balloon, it gets spread apart.
2. A hydrogen-filled balloon would spread the sound even more because it is even less dense than helium.
3. No. If you are in the water, the sound will simply bend, but not focus because the surface of the water is flat. Although sounds are different underwater, it is not because of this effect.

LAB 42: REFRACTION OF SOUND

QUESTION

How does sound bend as it passes from one medium to another? How can this be used as a model of how light acts?

SAFETY

Dispose of any popped balloon parts immediately after the lab.

MATERIALS

Helium balloon, empty balloon, glass or plastic soda bottle, vinegar (~125 ml), baking soda (~50 ml)

PROCEDURE

When light travels from one medium (material) to another medium with a different index of refraction, the light bends. This property can be used to focus light rays or spread them apart. The same thing happens with sound, but it is the density of the mediums that must be different.

1. Obtain a helium balloon. If you cannot find one, your teacher will have one available in class. Use a standard rubber balloon, not a metallic mylar balloon.
2. Fill a balloon with carbon dioxide. To do so, fill a small soda bottle with vinegar ¼ full. Roll up a medium handful of baking soda in a paper towel and twist it to fit through the neck of the bottle. Drop the baking soda into the bottle, remove the paper towel, and quickly put the balloon over the mouth. Wait for the balloon to stop inflating and tie it off.

   NOTE: SAVE YOUR CO_2-FILLED BALLOON FOR LAB 43.

3. Set up a radio or television and stand so that one ear is facing its speaker. Hold the helium balloon 30 cm from the ear facing the sound source and record your observations. Then hold the carbon dioxide balloon 30 cm from the ear and record your observations.

Post-Lab Questions

1. Helium is less dense than air. Carbon dioxide is more dense than air. When sound travels from air through the CO_2-filled balloon, it gets ______ (focused/spread apart). When sound travels from air through the helium balloon, it gets ______ (focused/spread apart).
2. How would you expect a hydrogen-filled balloon to sound?
3. Does this help explain how it sounds when you're underwater and someone on the surface is yelling to you? Explain why or why not.

LAB 43: BALLOONS AND RAY DIAGRAMS

This inquiry activity will be students' first exposure to ray diagrams. They will be using the refraction of sound to simulate the refraction of light to introduce them to this concept. They will be creating a simulated convex lens with a CO_2-filled balloon. A helium balloon is made optional because some students may not be able to purchase one and it cannot be easily included in the kit.

Answers to Procedure Questions

1. The sound should not change much because it is an air-filled balloon in an air-filled room. Because the air inside the balloon is compressed slightly, the sound may be a very small amount louder.
2. The sound should get louder as the CO_2 balloon focuses the sound on the ear.
3. The sound will get softer with a helium balloon.

Post-Lab Answers

1. Diagrams should look something like the one below:

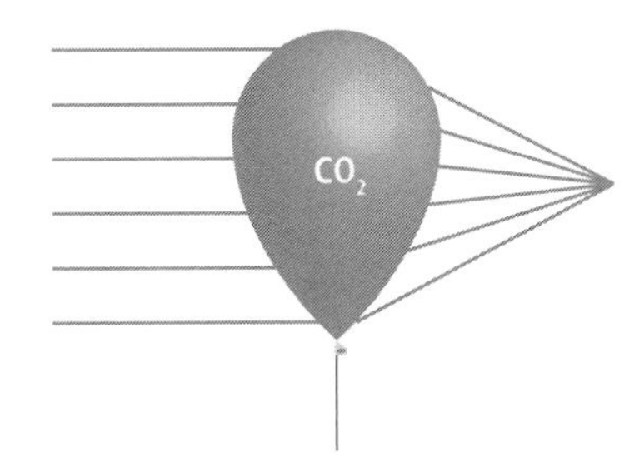

2. It was converging because the sound was focused, not spread out.

Section 3

3. Helium is less dense than air, so the opposite would happen; the rays would spread and the sound would be softer.

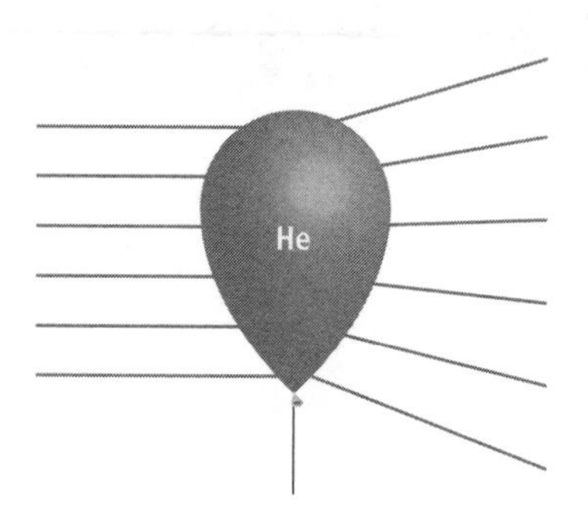

Note that the shape of the balloon affects the sound as it passes through. Because all the balloons are the same shape in this activity, the density of the air in the balloon is what makes the difference here.

LAB 43: BALLOONS AND RAY DIAGRAMS

QUESTION

How can you model the diffraction of sound with diagrams on paper?

SAFETY

Balloons are a choking hazard for young children.

MATERIALS

2 balloons; glass or plastic soda bottle; 1 helium balloon (optional); radio, television, loud sibling, or other source of sound; vinegar (~125 ml); baking soda (~50 ml).

PROCEDURE

When a sound wave passes from one medium to another medium with a different density, its speed changes. If the boundary between the two is curved or the wave enters at an angle, it can be refracted or bent. Which direction it bends depends on whether it is going from low density to high or high density to low.

Common lenses refract light because light has a lower speed through glass than through air. So a lens that is thicker in the middle (convex lens) bends light toward the middle of the lens (see Figure 43.1, p. 182). A lens that is thinner in the middle (concave lens) bends light away from the middle of the lens (see Figure 43.2, p. 182).

Sound can be bent the same way by passing it through materials with different densities. If the object acts like a converging lens, it can focus the sound and make it louder. If the object acts like a diverging lens, it will spread out the sound and make it softer.

You will be comparing two balloons: one filled with air and one filled with carbon dioxide.

Figure 43.1

Convex Lens (Converging)

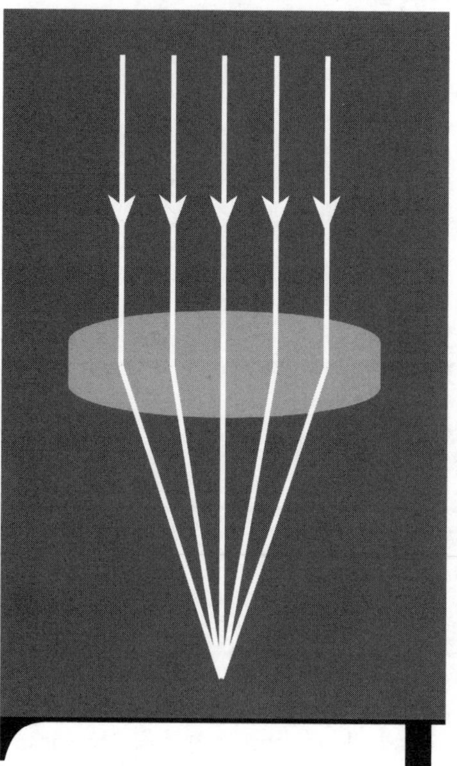

Concave Lens (Diverging)

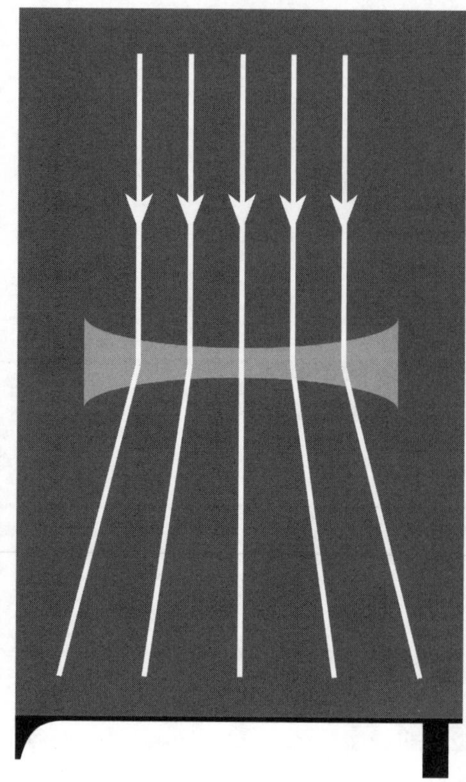

1. Set up a television or radio to be the source of sound for this experiment.
2. Blow up one of the balloons in your kit to a medium size.
3. If you saved your CO_2-filled balloon from Lab 42, use it for this step. If not, make a balloon by following these instructions: Fill a balloon with carbon dioxide. To do so, fill a small soda bottle with vinegar ¼ full. Roll up a medium handful of baking soda in a paper towel and twist it to fit through the neck of the bottle. Drop the baking soda into the bottle, remove the paper towel, and quickly put the balloon over the mouth. Wait for the balloon to stop inflating and tie it off.
4. Stand several feet away from the television or radio and listen to the sound. Now hold the air-filled balloon about 30 cm from your ear. Did the sound get louder, softer, or stay the same? Record your observations in a table or chart.
5. Repeat Step 5 with the carbon dioxide–filled balloon. Did the sound get louder, softer, or stay the same?
6. Predict what you think would happen if you did this experiment with a helium balloon. Would the sound get louder, softer, or stay the same?

Post-Lab Questions

1. Draw diagrams similar to those above (ray diagrams) to show what you think was going on in the balloons.
2. Was the carbon dioxide balloon a converging or a diverging lens?
3. What do you hypothesize would happen if you repeated this experiment with a helium balloon? Draw a ray diagram to explain your hypothesis.

Extensions

Try this experiment with a helium balloon and see if your prediction in #3 of the Post-Lab Questions was correct. Consider bringing the balloon to school so that another student without one can experience it as well. Some science museums have giant balloons that focus sounds from the other side. Take a trip to one of these museums and experience this sensation on a large scale.

LAB 44: LENSES AND RAY DIAGRAMS

This is the first time that students will draw technical ray diagrams (in previous experiments, they simply sketched the rays). They need a little instruction in drawing ray diagrams before they can do it, so the activity is not purely inquiry. But it is inquiry in learning about images and their properties. Students will see what it means to be reduced or enlarged, erect or inverted, real or virtual.

Topic: Lenses
Go to: *www.sciLINKS.org*
Code: THP26

Post-Lab Answers

1. If an object is beyond the focal point of the lens, the image will be reduced and inverted.
2. If an object is inside the focal point of the lens, the image will be enlarged and erect.
3. Answers will vary but should be less than 6 cm. The short focal length helps give it more magnification.

LAB 44: LENSES AND RAY DIAGRAMS

QUESTION

How does an object look through a magnifying lens at different distances?

SAFETY

Some magnifying glasses are made of glass; be careful if the glass should break.

MATERIALS

Magnifying glass

PROCEDURE

Lenses refract light because light travels at a lower speed through glass than through air. So a lens that is thicker in the middle (convex lens) bends light toward the middle of the lens (see Figure 44.1, p. 187). A lens that is thinner in the middle (concave lens) bends light away from the middle of the lens. A magnifying glass is simply a convex lens.

1. Hold the magnifying glass out at arm's length and look at something far away (but not the Sun). Do the objects that you see through the lens look right-side up (erect) or upside down (inverted)? Big or small?
2. See the ray diagram on page 188 (Figure 44.2) as a reference for ray diagrams that you will draw in the future. Rays are drawn such that a ray that comes in parallel to the horizontal will bend through the focal point and keep going. A ray that passes straight through the center of the lens will not bend at all. Where these two lines cross is where the image will appear and will show you how big the image will appear and whether it will be erect or inverted. If the rays cross below the horizontal, it will be inverted. If they cross above the line, they will be erect.

Figure 44.1

Convex Lens (Converging)

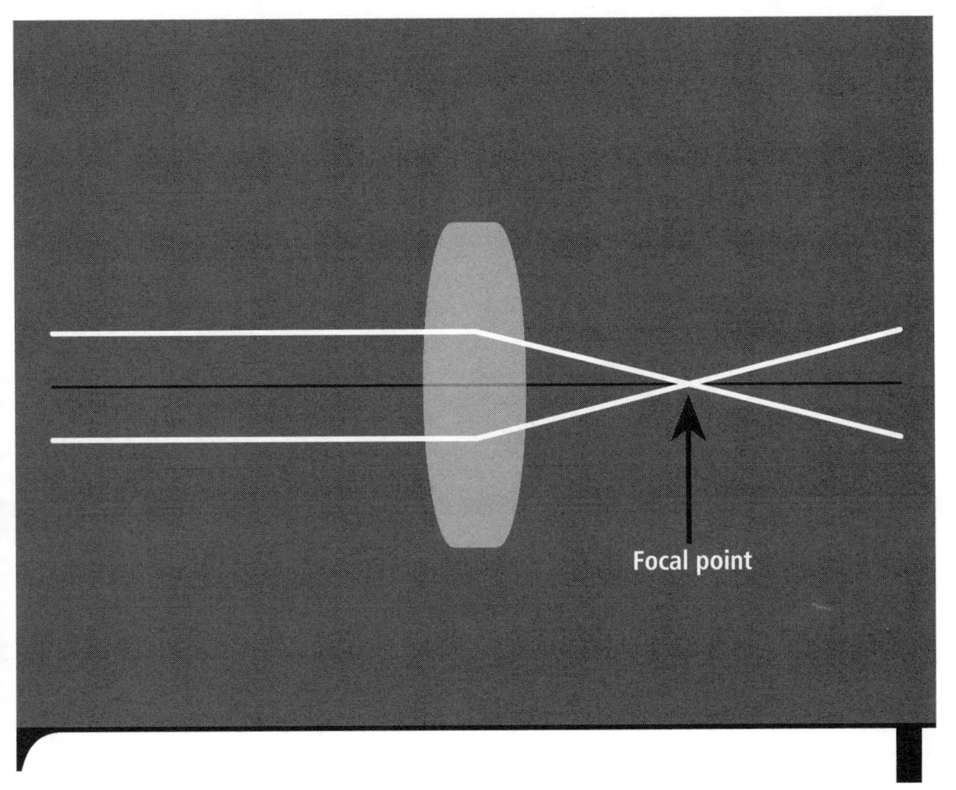

3. Now hold the magnifying glass at arm's length and look at something only an inch away. Do the objects that you see through the lens look right-side up (erect) or upside down (inverted)? Big or small?
4. Draw a ray diagram similar to the one above but put the object close to the lens. Be sure that the object is closer to the lens than the focal point. Does the result confirm your answers in Step 3?
5. Hold the object at arm's length and move it back and forth away from and toward an object until you find the point between right-side up and upside down where the image gets blurry. The distance from the lens to the object in this case is the focal length because the object is right at the focal point. Measure this distance. This is the distance used to burn objects with a magnifying glass because all the light is focused on one spot.

Post-Lab Questions

1. Finish this statement: If an object is beyond the focal point of a convex lens, the image will be __________ (enlarged/reduced) and __________ (inverted/erect).

Figure 44.2

Ray Diagram

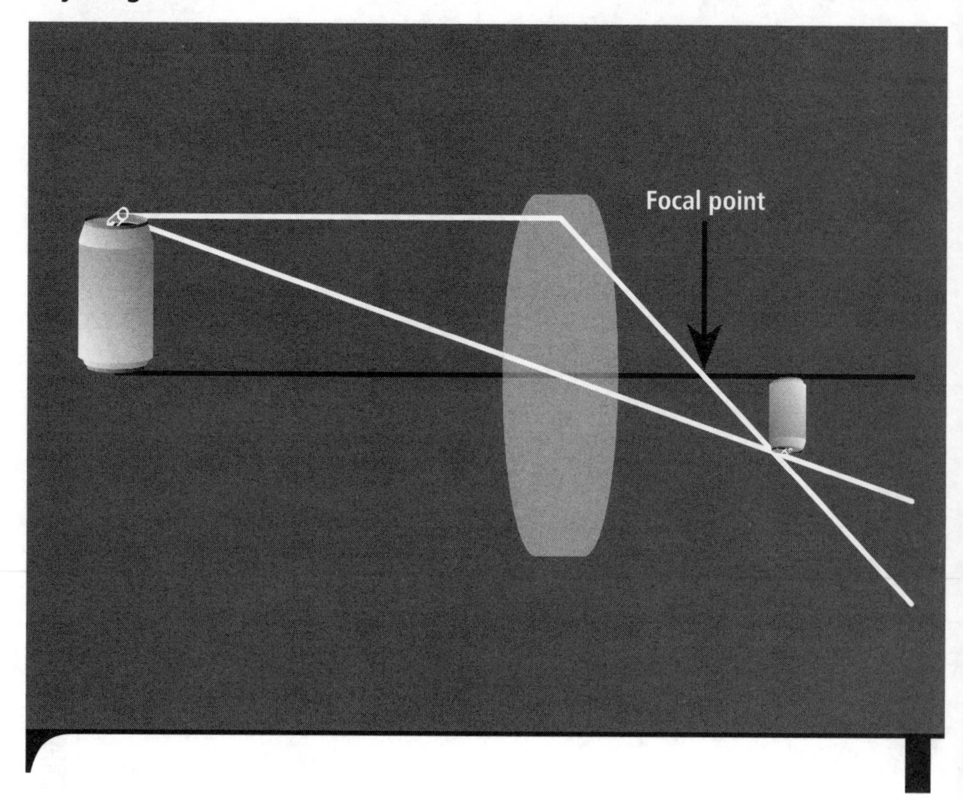

2. Finish this statement: If an object is inside the focal point of a convex lens, the image will be __________ (enlarged/reduced) and ___________ (inverted/erect).
3. What was the focal length of your lens?

Extension

If you have access to other magnifying glasses, see if you can determine how the magnification of the lens is related to the focal length. See if you can draw ray diagrams to verify your conclusion.

LAB 45: CURVED MIRRORS

This inquiry activity will be used before discussing curved mirrors in class. Students will discover how curved mirrors act and how the size and the orientation of the image are related to the distance from the mirror. Ray diagrams for curved mirrors are more complicated than ray diagrams for lenses, so this lab is not an attempt to get students to draw ray diagrams. The lab uses the term *focal point* loosely. Parabolic mirrors have true focal points; spoons do not. The lack of a true focal point is demonstrated by the distortions in the spoon reflections. However, the terminology is convenient for our purposes, as the spoons have a focal point–like area.

Topic: Mirrors
Go to: *www.sciLINKS.org*
Code: THP27

Post-Lab Answers

1. If an object is beyond the focal point of the concave mirror, the image will be reduced and inverted.
2. If an object is inside the focal point of the concave mirror, the image will be enlarged and erect.
3. If an object is beyond the focal point of the convex mirror, the image will be reduced and erect.
4. If an object is inside the focal point of the convex mirror, the image will be reduced and erect.
5. Answers will vary based on the curvature of the spoon.

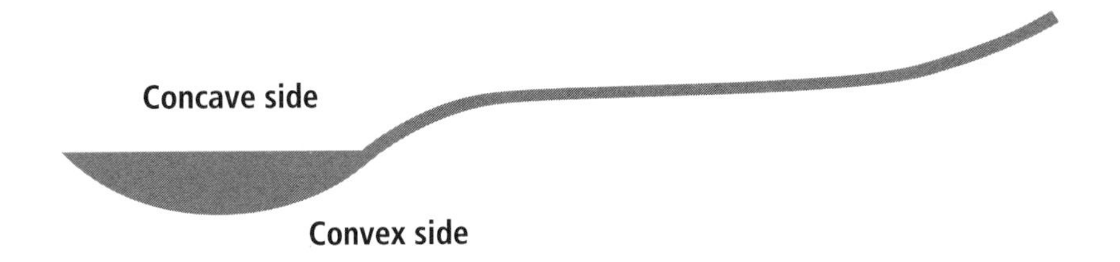

LAB 45: CURVED MIRRORS

QUESTION

What do objects look like in a curved mirror at different distances?

SAFETY

Standard safety precautions apply.

MATERIALS

Large, shiny metal spoon

PROCEDURE

Mirrors reflect light. Flat (plane) mirrors reflect light rays out at the same angle that they come in (see Figure 45.1, p. 191). Curved mirrors can appear to violate this if the normal line is not drawn perpendicular to the mirror (see Figure 45.2, p. 191).

When a mirror is curved, the shape of the surface changes the direction of the reflected light in ways that can be useful. Converging mirrors (concave) have a focal point like a lens does through which they focus the light. Diverging mirrors (convex) focus light away from the focal point on the back side of the mirror. For a concave mirror, one ray is drawn parallel to the horizontal axis from the top of the object, and it is reflected through the focal point (solid line in Figure 45.3, p. 191). A second ray is drawn from the top of the object through the focal point, and it is reflected out parallel to the horizontal (dashed line in Figure 45.3). Where they cross, the image will form. If the object is inside the focal point, then the second ray comes out of the focal point, through the top of the object, and bounces off the mirror parallel to the horizontal.

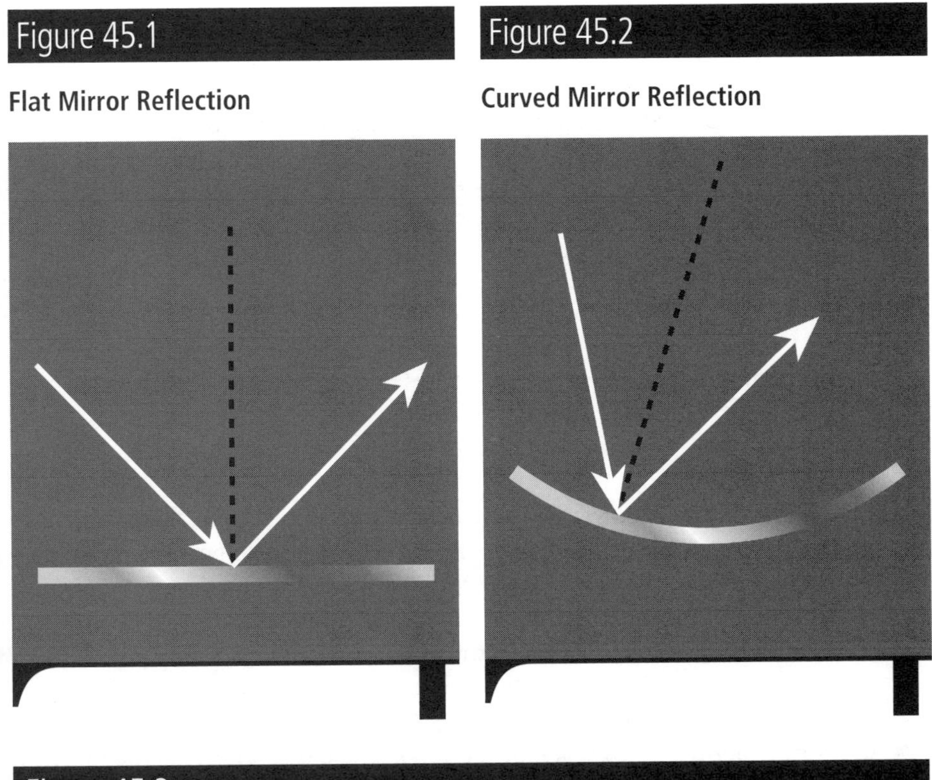

Figure 45.1

Flat Mirror Reflection

Figure 45.2

Curved Mirror Reflection

Figure 45.3

Ray Diagram for a Concave Mirror

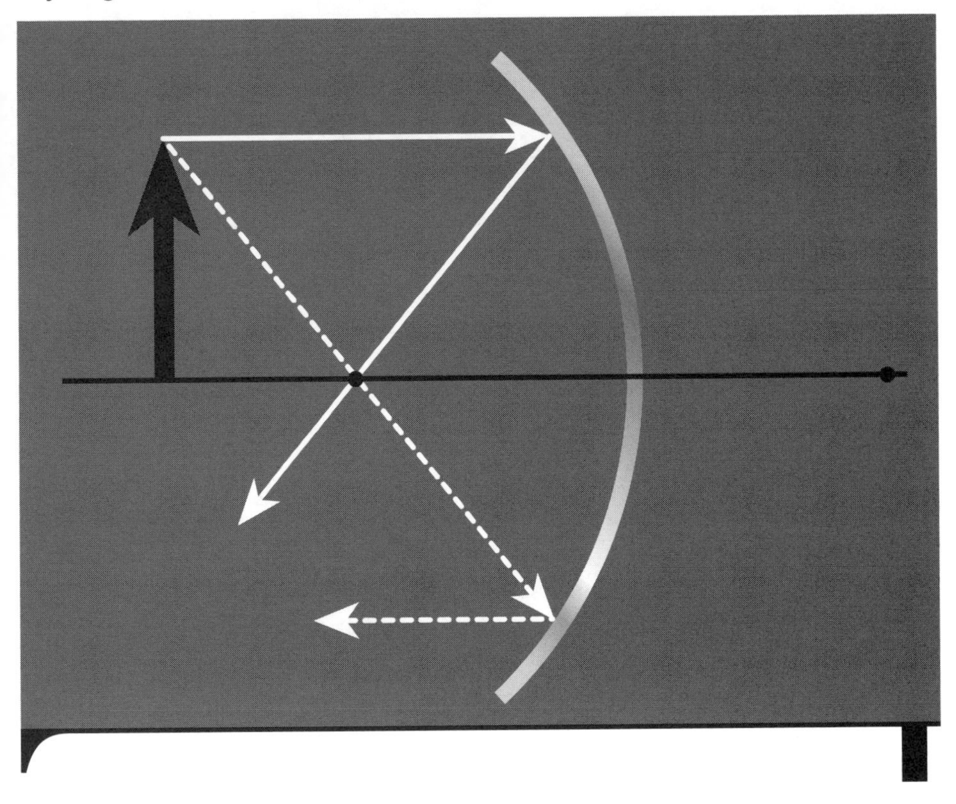

In this activity, you will not be expected to draw ray diagrams for mirrors; you will have to learn and practice that with your teacher. But you will discover how a curved mirror acts and then you will be able to explain those behaviors once you know how to draw the ray diagrams.

1. Hold the spoon so that the concave side is facing you. Hold it far away from your face and look at the image of your face. Is it large or small? Right-side up or upside down?

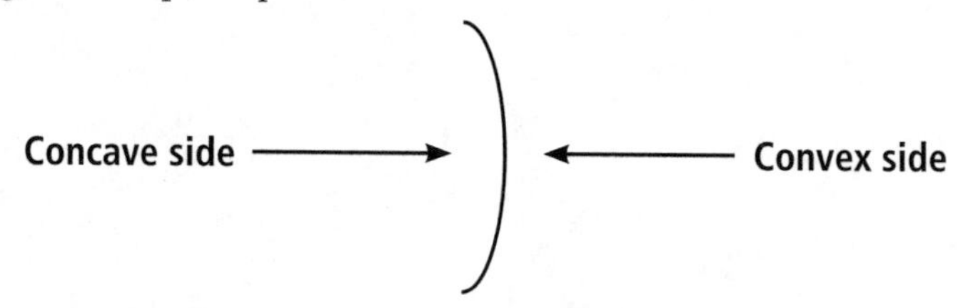

2. Now hold the spoon with the concave side facing you and move it very close to one of your eyes. Is the image of your eye larger or smaller than normal? Right-side up or upside down? You can move the spoon in and out and up and down if it helps you answer these questions.
3. Now flip the spoon around so that the convex side is facing you. Hold it far away from your face. Is the image of your face large or small? Right-side up or upside down?
4. Now bring the convex side of the spoon close to your eye. Is the image of your face large or small? Right-side up or upside down?
5. Come up with your own technique for figuring out where the focal point of the spoon is and measure the focal length.

Post-Lab Questions

1. Finish this statement: If an object is beyond the focal point of the concave mirror, the image will be __________ (enlarged/reduced) and __________ (inverted/erect).
2. Finish this statement: If an object is inside the focal point of the concave mirror, the image will be __________ (enlarged/reduced) and __________ (inverted/erect).
3. Finish this statement: If an object is beyond the focal point of the convex mirror, the image will be __________ (enlarged/reduced) and __________ (inverted/erect).
4. Finish this statement: If an object is inside the focal point of the convex mirror, the image will be __________ (enlarged/reduced) and __________ (inverted/erect).
5. What was the focal length of your spoon mirror?

Extension

Why do some gas stations or convenience markets have large curved mirrors in the corner of the store? How does the curved mirror help them?

LAB 46: COLOR ADDITION

This activity is inquiry in that students do not know how colors are combined. They likely think that the primary colors are red, yellow, and blue. In fact, there are two sets of primary colors: red, green, and blue for additive colors of light, and cyan, magenta, and yellow for subtractive colors of pigments.

It is more difficult to teach the subtractive colors at home, so the teacher will have to do that in the classroom.

SCILINKS. THE WORLD'S A CLICK AWAY
Topic: Color
Go to: *www.sciLINKS.org*
Code: THP28

Post-Lab Answers

1. Red and green made yellow. It will probably be surprising. Red and blue make magenta. Green and blue make cyan. It is not coincidence that the additive primaries mix to make the subtractive primaries. The subtractive primaries mix to make the additive primaries.
2. RGB stands for red, green, blue. Those three colors of light can make any color of the rainbow.
3. White is 100% red, 100% blue, and 100% green. By varying the percentages, you can get intermediate colors such as pink, brown, and gray.

LAB 46: COLOR ADDITION

QUESTION

What are the primary additive colors and how are they used to create all of the colors on a computer screen?

SAFETY

Be careful with water near any electronic devices.

MATERIALS

Magnifying glass, computer monitor

PROCEDURE

Normally, in elementary school, you are taught that the primary colors are red, yellow, and blue. There are two sets of primary colors (additive and subtractive) and neither set is red, yellow, and blue. In this lab, you will see how additive primary colors are used to create the spectrum of colors on a computer monitor. Note that the primary colors are only primary because of the way that the human eye works, not because of a law of physics. We have red, green, and blue receptors in our eye—therefore those are the primary colors for humans.

1. Put a picture on your computer monitor that contains the colors black, white, yellow, and green.
2. Use your magnifying glass to look closely at an area that is white. You should see that there are groups of dots (called pixels). Each pixel is made of three small dots of colors (called sub-pixels) that make up the color white. What are they? These are the primary additive colors and are in different proportions to make all of the other colors.

3. Use the magnifying glass for an area that is black. Which colored sub-pixels do you see? Record your answer.
4. Dip your finger in water and touch an area of the screen that is red. Which sub-pixels do you see? Record your answer.
5. Repeat Step 4 with yellow and blue.

Post-Lab Questions

1. What combination of colored sub-pixels made yellow? Did this surprise you? What combination of additive primaries would it take to make magenta? To make cyan?
2. When you plug a computer into a multimedia projector, you plug it into the RGB Input port. What does RGB stand for?
3. How does the computer monitor create all the other thousands or millions of colors from just these three colored pixels? Use the magnifying glass to test your theory.

LAB 47: DIAMETER OF THE SUN

Although this lab is not an inquiry activity, it is another exercise to help students think about light rays and ray diagrams. The students think that the point of the lab is to calculate the diameter of the Sun. Really, the point is to get used to thinking about light as rays to better understand light passing through a lens or bouncing off a mirror. Students will start to think about erect and inverted images as a result as well.

Some wonder how this lab works considering that it is said that the light rays coming from the Sun travel parallel and not at an angle. It is true that most of the rays travel parallel, but the ones coming from the edges of the Sun to us vary by about 0.5 degrees from parallel.

Post-Lab Answers

1. 1,400,000,000 m or about 110 Earth diameters. Percentage errors should be less than 20% and are more likely less than 10%.
2. Because the rays from the top of the Sun will be on the bottom of the image, the image will be inverted (upside down).
3. Christian Huygens used the angle between the Sun, Earth, and Venus and geometry to figure it out. He used Venus because it goes through phases like Earth's moon (crescent, full, new, etc.), and this allowed him to determine said angle. He actually guessed the diameter of Venus and coincidentally got very close.

LAB 47: DIAMETER OF THE SUN

QUESTION

How can you determine the diameter of the Sun with simple materials?

SAFETY

Do not look directly at the Sun.

MATERIALS

3 × 5 in. card, hole punch, ruler, penny

PROCEDURE

Early scientists were able to calculate seemingly impossible measurements, including the distance to the Sun and the circumference of the Earth. In this lab, you will use a 3 × 5 in. card with a hole punched in it to calculate the diameter of the Sun. You will use an inverse ratio based on similar triangles to figure it out.

1. If there is not already a hole in your 3 × 5 in. card, punch a hole in it using a regular hole punch.
2. Measure the diameter of the penny in centimeters and convert it to meters.
3. Go outside near noon, when the Sun is highest in the sky, and hold out your card over level ground about waist high. Notice that the Sun is showing through the hole in the card.
4. Move the card up and down and notice that the circle gets bigger and smaller.
5. If your spot of light is not a circle because the Sun is not straight up, you can project it onto a wall or a book slanted at an angle until it is circular.

6. Move the card up and down until it just covers the penny, and then measure the distance from the penny to the card in meters.

Figure 47.1

Finding the Diameter of the Sun

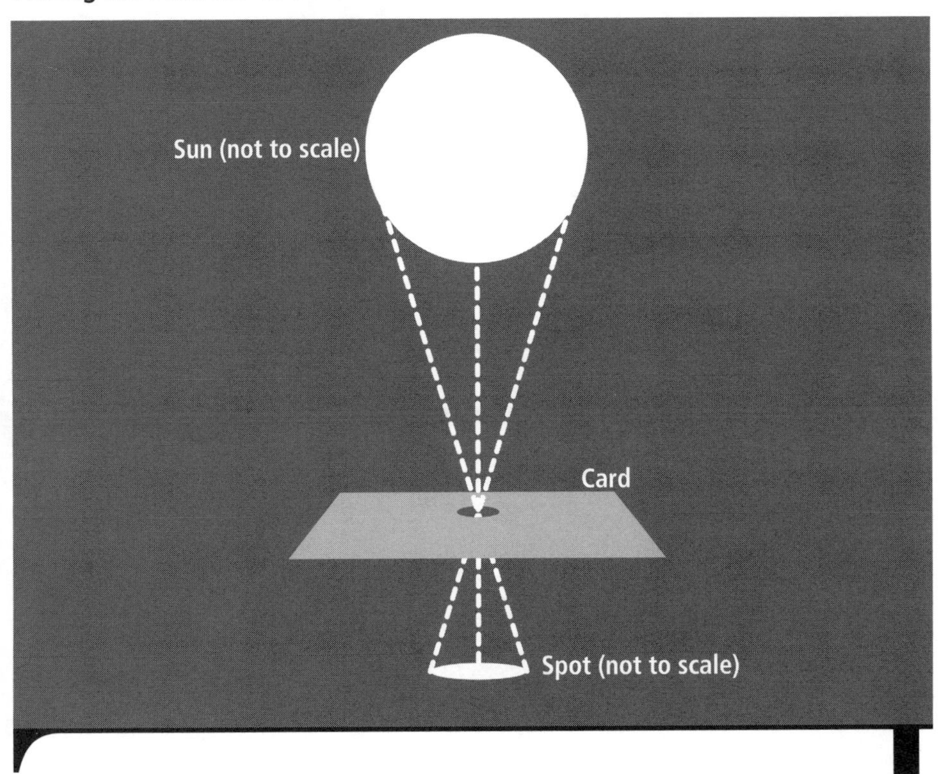

7. In Figure 47.1 above, the two triangles that meet near the hole in the card have the same angle (they are vertical angles), so the triangles are similar triangles. Because of this, the corresponding parts of the triangles are in the same ratio. Therefore, the diameter of the spot divided by the distance from the card to the spot is equal to the diameter of the Sun divided by the distance to the Sun.

$$\frac{\text{Diameter of spot}}{\text{Distance to spot}} = \frac{\text{Diameter of Sun}}{\text{Distance to Sun}}$$

8. The distance to the Sun is 150×10^9 m. Calculate the diameter of the Sun.

Post-Lab Questions

1. Check an astronomy book or the internet and calculate your percentage error.

2. If the Sun were not a sphere, would the image you saw be erect (right-side up) or inverted (upside down)?
3. Use a book or the internet and find out how ancient scientists knew how far away the Sun is.

LAB 48: INTENSITY VERSUS DISTANCE

This activity is Level 2 inquiry in that students will not know in advance that the relationship is inverse square. Although they may not get a perfect inverse square graph, it will be clear that it is not a perfect straight line. This activity can also be done on a smaller scale on graph paper with a smaller cardboard cutout if there is not sufficient wall space.

Topic: Properties of Light
Go to: *www.sciLINKS.org*
Code: THP29

Post-Lab Answers

1. It would have been a square 3 cm on a side (9 cm^2) because at 0 ft. it would be touching the wall and we cut out a 9 cm^2 square.
2. Answers will vary.
3. Answers will vary but should resemble an inverse square graph.
4. Inverse square
5. 64 cm

$$\frac{I_1}{I_2} = \frac{\frac{1}{(x_1)^2}}{\frac{1}{(x_2)^2}} \quad \text{so, } 2 = (1/x^2)/(1.23 \times 10^{-4}) \quad 2.46 \times 10^{-4} = 1/x^2 \quad x = 64 \text{ cm}$$

6. No, it was ¼ the intensity. No, it was ½ the intensity.
7. Answers will vary.

LAB 48: INTENSITY VERSUS DISTANCE

QUESTION

How does the intensity of a light change as the light moves farther away?

SAFETY

Look around the room before turning off the lights to note things that could be tripped over.

MATERIALS

Flashlight, bare wall, 3 cm square cutout from thick paper, ruler, graph paper

PROCEDURE

In this lab, you will be shining a flashlight on a wall and measuring the area that the light covers. Because the amount of light coming out of the flashlight does not change as you walk closer to and farther from the wall, the area that it covers can be related to how intense the light is that is hitting the wall. The larger the area that the light is spread over, the lower the intensity. You will be using a graph to determine what type of relationship exists between intensity of a light and distance from the source.

1. Cut out a 3 cm square from a thick piece of paper or cardboard.
2. Put objects on the floor to mark 30, 60, 90, 120, 150, and 180 cm away from a bare wall in a fairly dark room.
3. Have an assistant hold the 3 cm square in front of the flashlight and hold it 30 cm from the wall and measure the size of the square on the wall. Only measure the bright, distinct shape, not the dim shape around it (see Figure 48.1, p. 203).

Figure 48.1

Measuring the Area of Light

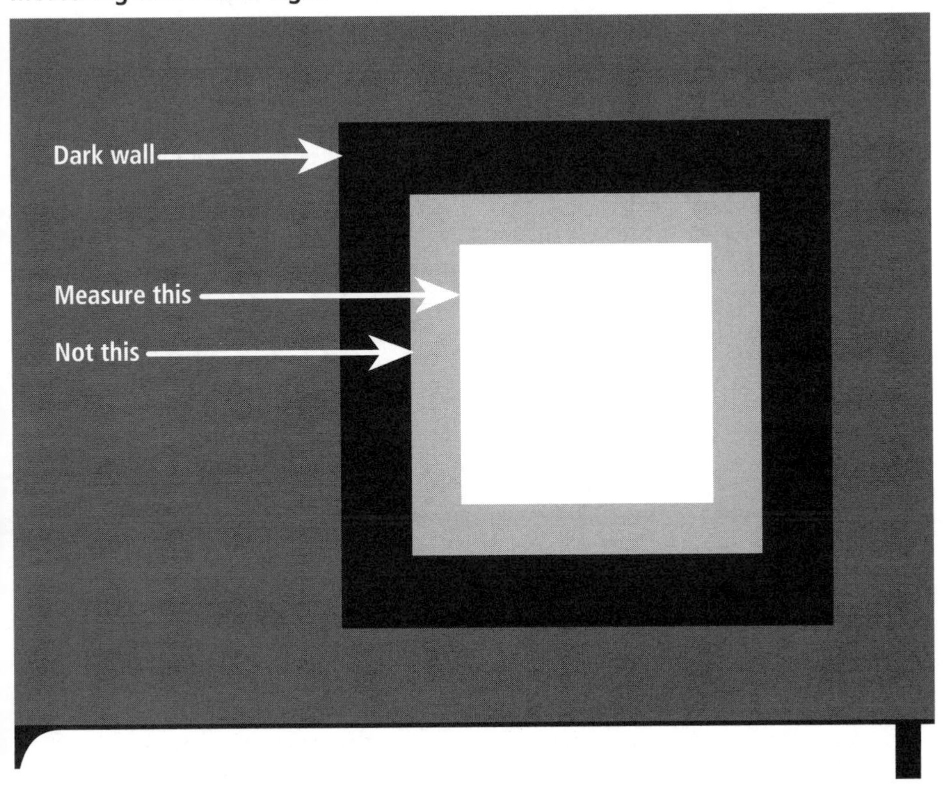

4. Repeat this procedure at 60, 90, 120, 150, and 180 cm away and record your data in a data chart like the one below.

Data Chart

Distance	0 cm	30 cm	60 cm	90 cm	120 cm	150 cm
Width of Square						
Intensity		See Post-Lab Question 2 for intensity calculation.				

Post-Lab Questions

1. What would have been the size of the square if the flashlight were at 0 ft. (touching the wall)? Put that in your data chart as well. *Hint:* The answer is *not* 0 cm.
2. Call the amount of light coming out of the flashlight 1,000 light units. For each data point, calculate the light/square cm (intensity) and record it in your data chart. Assume that the square has equal height and width.
3. Draw a graph with distance on the horizontal axis and "light per square cm" on the vertical axis.
4. The shape of the graph indicates a(n) ________ (direct, inverse, square, inverse square) relationship between distance from a light source and light intensity.

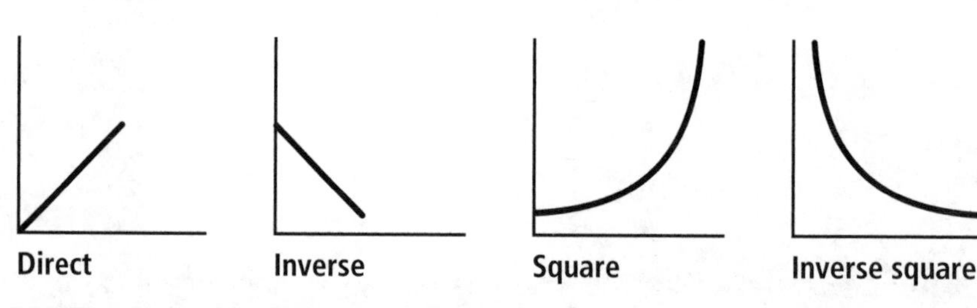

5. How far away would you have to be to get twice the light intensity that you had at 90 cm?
6. When you were at 120 cm, was the intensity half as much as when you were at 60 cm? If not, what fraction was it? At 90 cm, was the intensity one-third as much as when you were at 30 cm? If not, what fraction was it?
7. Do you think the graph would have been any different had you used a circle instead of a square? Explain your answer.

Extension

The Sun is a distance of one Astronomical Unit (AU) away from the Earth. If there was a star that was exactly the same size and brightness as the Sun but was a million AU from the Earth, what would be its intensity relative to the Sun?

LAB 49: RIPPLE TANK

In this inquiry activity, students will discover some of the rules of reflection of water waves, how waves travel around boundaries, and how waves move floating objects before they have learned about these concepts. If students have difficulty seeing the waves, adjusting the angle of the ripple tank to the incident light makes a big difference. Food coloring can also make a difference.

Topic: Nature of Waves
Go to: *www.sciLINKS.org*
Code: THP30

Post-Lab Answers

1. The wave got higher. As the width of the wave decreased, the amplitude increased. The energy and the water were concentrated into a smaller space. The beach shaped like a wedge is probably so popular because the waves are bigger than other areas.
2. They are a combination of the two. The floating object traveled in a circular pattern, which is a combination of back-and-forth and up-and-down motion.
3. No. Once the standing wave is originated, it does not take much energy to keep it going. It also does not take much energy to keep a standing wave going in a slinky. It is so easy to create a standing wave in air that you can do it by blowing across the top of a bottle.

LAB 49: RIPPLE TANK

QUESTION

How do barriers affect water waves?

SAFETY

Standard safety precautions apply.

MATERIALS

Plastic box, water, CD case, food coloring (optional)

PROCEDURE

A ripple tank is a device used to study water waves. It is basically a tank with water in it in which the user creates waves. Then barriers can be placed at different locations to study how the waves reflect, refract, and interfere.

You will make a ripple tank out of the box that your lab materials came in. Adding some color to the water can make the waves easier to see. A plastic CD case can be used to cause the vibrations in the water.

1. Fill the plastic box with approximately 4 cm of water and add color to it if you wish.
2. Use the plastic CD case or a similar object to set up waves in the tank, and observe them.
   a. Do the waves appear to be longitudinal or transverse?
   b. If you take the CD case out once the waves are set up, how many times do the waves travel back and forth before stopping?
   c. Calculate the speed of the wave by timing how long it takes to go back and forth several times. Calculate the total distance that the wave traveled and divide by the time that it took.

3. Move the CD case back and forth in sync with the waves until the wave appears to stay in place. This is called a standing wave.
   a. Calculate the period of the standing wave.
   b. Calculate the frequency of the standing wave.

4. Find two objects that almost reach across the box, but leave a small gap of about 3 cm in the middle. You can use small plastic rulers, pieces of wood, clay, etc. Have someone help you hold the objects while you create waves that will hit the barriers. Be sure that the water is not deeper than the barriers are tall. Draw the waves as they pass through the gap in the barrier below. When waves pass through a small opening, they are diffracted.

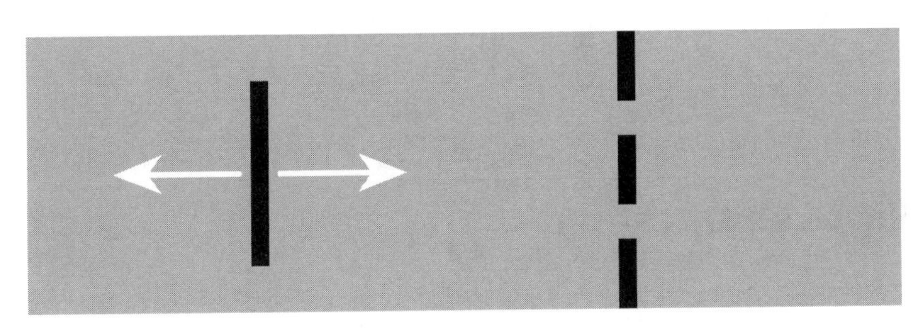

5. Now find three objects that will leave two gaps as shown below. Have your assistant help you again while you create waves to hit the barriers. This time, waves will be diffracted through both gaps and then will interfere on the other side. Sketch the interference pattern that you see.

6. Now place the two objects so that they form a wedge with a gap in the middle as shown below. Make observations about the waves as they approach the gap and after they pass through. Make observations from the side as well.

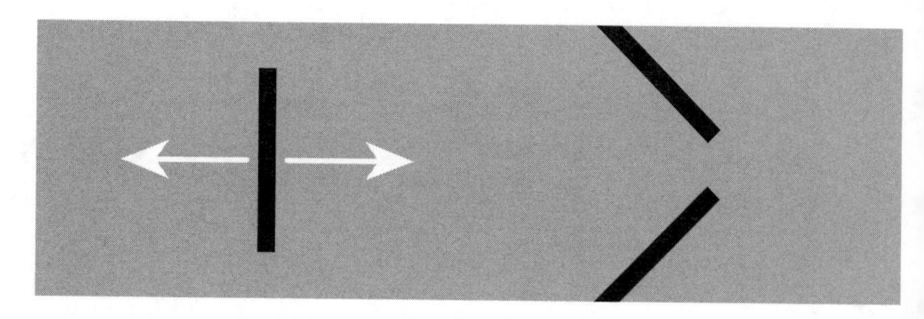

7. Put a floating object in the water and watch from the side as you create a standing wave. Sketch the path that the object follows.

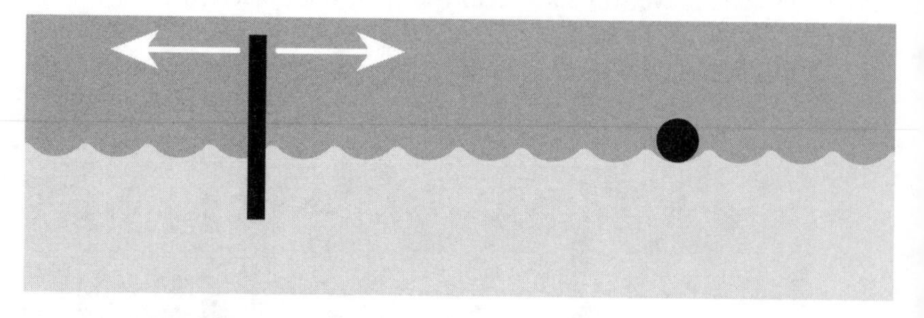

Post-Lab Questions

1. How did the wedge affect the amplitude (height) of the wave? There is a popular surfing destination in California called The Wedge where the rocks form a wedge. Why do you think this is such a popular destination?
2. Based on the path that the floating object followed in Step 7, would you say that water waves are transverse (move up and down), longitudinal (move back and forth), or a combination of the two?
3. Did it take a lot of effort to keep the standing wave going? Does it take a lot of effort to keep a standing wave going on a slinky?

Extension

Do research into interference of light when passing through a single slit, double slit, or diffraction grating. How does the diffraction of light through slits compare to what you saw when water waves passed through slits?

LAB 50: OIL SPOT PHOTOMETER

Students will once again be reminded of the inverse square nature of light intensity. This is a concept that students often struggle with, especially when calculations are involved. The important idea of this lab is not to actually figure out the wattage of a bulb, but to practice using the inverse square law.

Post-Lab Answers

1. Answers will vary, but they should be close. The closer the spot is to the bulb, the more accurate the technique should be. The brighter light gives more resolution when looking at the spot. But too close and the flashlight will not be bright enough to balance it out.
2. One is that the observer cannot tell exactly when the spot is invisible. There are a range of distances where it is approximately invisible. The other major source of error is the lamp. If there is not a good reflector on the lamp and there is a good reflector on the flashlight, then a higher percentage of the flashlight's power will hit the oil spot than the lamp's, and the comparison won't be even.
3. It certainly could, but the reflector issue in #2 would become huge. Most of the Sun's power never hits Earth, so the experiment would only be figuring out the wattage that strikes ~4 cm^2 of the Earth's surface.
4. Yes. Because both sides are subject to an inverse square relationship, the relationship cancels out. Assume that the lightbulb is 100 watts and the flashlight is 1 watt. Calculate the distance that the flashlight would have to be if the lamp is at 1 m and 2 m.

$$\frac{100}{1^2} = \frac{1}{x^2} \qquad x = 0.1 \text{ m}$$

$$\frac{100}{2^2} = \frac{1}{x^2} \qquad x = 0.2 \text{ m}$$

LAB 50: OIL SPOT PHOTOMETER

QUESTION

What is the wattage of a flashlight bulb?

SAFETY

Be careful when dealing with the lamp. The heat from the bulb and the electricity both pose a danger. Put the lamp back when you're finished.

MATERIALS

Blank paper, cooking oil, flashlight, lamp, calculator

PROCEDURE

A photometer is a device that measures the brightness of a source of light. Technically, the instrument measures the "luminosity" of the light source. There are very accurate electronic photometers that can be purchased, but you will be making one for only a few cents. The one that you will be making will tell you when the brightness of two lights is the same on either side of the photometer. If the wattage of one of the lightbulbs is known, then the other can be figured out using the following equation:

$$\frac{W_1}{(D_1)^2} = \frac{W_2}{(D_2)^2}$$

In this equation, W stands for the wattage (power) of the bulb and D stands for the distance between the bulb and the photometer. The units for both don't matter as long as they are the same on both sides. In this lab, you can use lumens or watts for W and centimeters or meters for D. The 1 and 2 stand for lightbulb 1 and lightbulb 2. The relationship between the brightness of the light and the distance from the light is called an inverse square relationship, as you discovered in Lab 48: Intensity Versus Distance.

1. To make the photometer, put a couple of drops of cooking oil in the center of a sheet of white paper and spread them into a circle approximately 2 cm in diameter. This oil spot will almost disappear when the brightness of the bulb on either side of it is the same. If two equal bulbs were used, they would have to be the same distance apart. But you will not be using two equal bulbs.
2. Locate a place where there is an incandescent lightbulb available and record its wattage or lumens if you still have the package. A table lamp with a reflector will be best. Find a flashlight and an assistant.
3. Place the photometer paper 50 cm from the lightbulb and hold it there. Have your assistant move the flashlight closer to and farther away from the photometer until the oil spot nearly disappears. Measure and record the distance between the photometer and the flashlight.
4. Repeat the experiment again but this time hold the paper 100 cm from the lightbulb and record the distance to the flashlight that makes it equal brightness.

Calculations

Use the equation in the Procedure introduction to calculate the wattage (power) of the flashlight when the light was 50 cm away. Repeat the calculation with your data from 100 cm away.

Post-Lab Questions

1. How did your answer for the wattage of the bulb for the two trials compare? Which would you think was more accurate? Explain why.
2. What are some sources of error involved in this lab?
3. Do you think that this could be used to find the power of the Sun? What other information would you need? What would be some additional sources of error?
4. Was the flashlight twice as far away when the lightbulb was twice as far away? Explain.

Extension

Many compact fluorescent bulb manufacturers claim that they can produce the same amount of light as a 75-watt incandescent bulb while using far less electricity. The wattage printed on the bulbs that you buy is how much electricity they use. See if a compact fluorescent produces more light than an equal wattage incandescent.

LAB 51: WAVES AND INTERFERENCE

This activity is inquiry in that students have not yet studied interference or interference patterns. Students without a computer at home should be instructed as to where they can go to use a computer with speakers that will allow them to install software on it. This software can also be used to demonstrate "beat frequencies" by changing the frequency in one speaker slightly compared to the other.

Use the next page to copy the circles onto transparency film and give each student one set of four circles. Be careful when creating transparencies as you can damage a printer if you do not use the correct film. There is special transparency material for ink jet printers and special material for dry toner printers (laser printers and photocopy machines). Do not mix up the two or the sheets could melt in your printer and cause irreparable damage.

Topic: Interference
Go to: *www.sciLINKS.org*
Code: THP31

Post-Lab Answers

1. As frequency gets higher, the loud spots get closer. The loud spots should be separated by one wavelength. So a frequency of 2,000 Hz should have a wavelength of 17 cm.
2. Graphs will vary based on the acoustics of the room and how carefully the student measures, but should approximate a straight line.
3. Graphs will vary, but using velocity = frequency (wavelength), a 20 Hz wave should have a wavelength of 17 m. You cannot experience that in a chair in a small room. A wave with a frequency of 20,000 Hz should have a wavelength of 1.7 cm, and that is likely too small to measure using this method.
4. The loudness does not have an effect on how far apart the loud spots are.

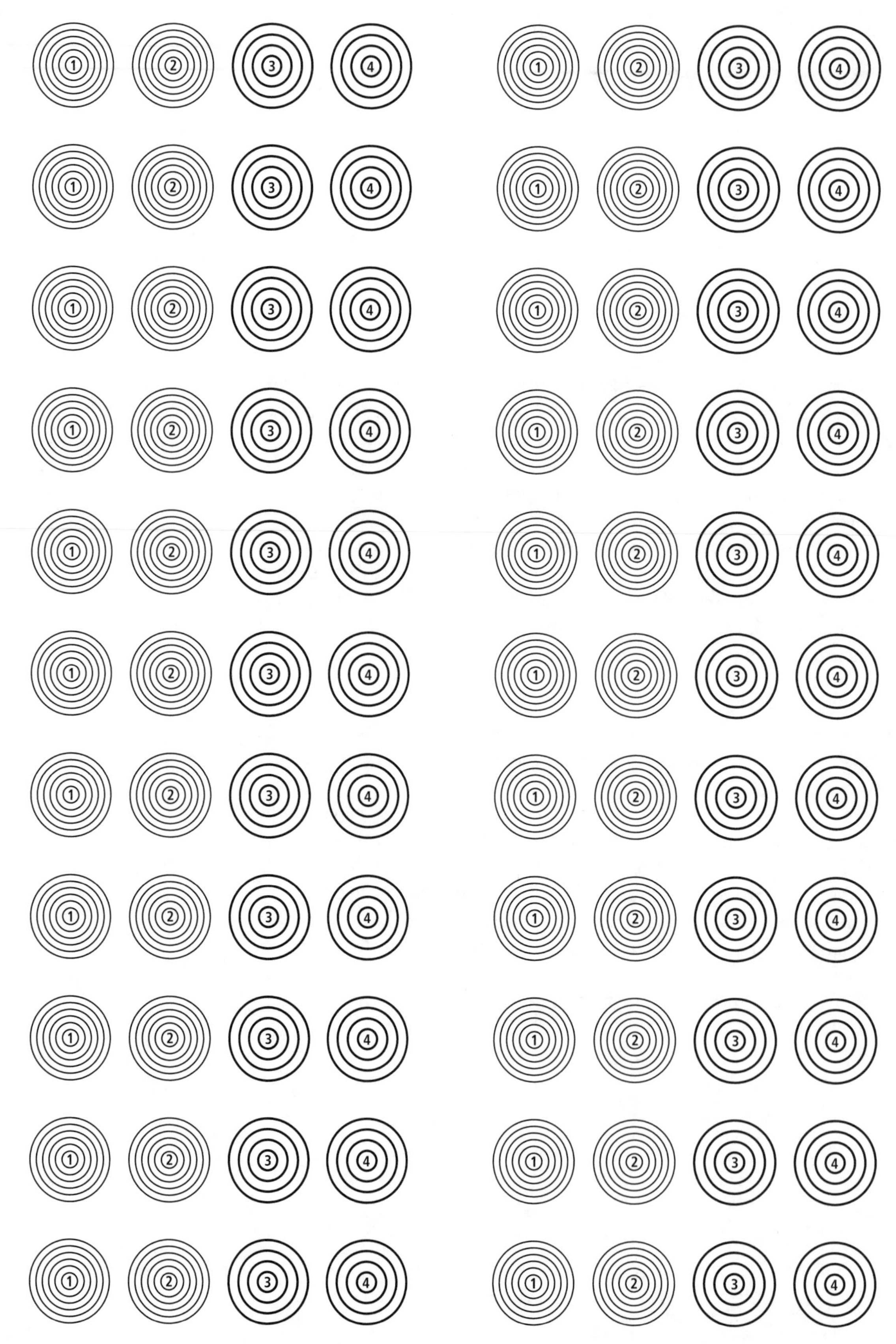

1
2
3
4

LAB 51: WAVES AND INTERFERENCE

QUESTION

Why are there loud spots and "dead" spots when music is played in large rooms?

SAFETY

Standard safety precautions apply.

MATERIALS

Concentric circles printed on transparency material, computer, tone generator program

PROCEDURE

When two waves of any kind cross each other's path, they interfere either constructively (increase amplitude) or destructively (decrease amplitude). When sound waves do this, the result can be heard. You will be simulating wave interference with some models of sound waves. Then you will download software on a computer that will allow you to experience the interference for yourself.

To do this, you will need a Windows computer with a sound card and stereo speakers. It will not work with headphones. Go to *www.nch.com.au/tonegen/index.html* to download the evaluation version of the NCH Tone Generator software and install it on your computer. If this website changes, simply search for "Free Tone Generator" on a search engine and several may be found.

Part 1

1. Cut out the four sets of concentric circles on the transparency. These circles represent sound waves emanating from a source at the center. The black parts can be thought of as areas of compression and the clear areas are areas of rarefaction.

2. Put circles 1 and 2 so that they overlap about halfway.
   a. What do areas where black crosses black represent? Are these areas loud or soft?
   b. What do areas where white and white cross represent? Are these loud or soft?
   c. What do areas where white and black cross represent? Are these loud or soft?
   d. Measure how far apart two consecutive dark crossing spots are. __________ cm
3. Put circles 3 and 4 so that they overlap about halfway.
   a. Would these sounds be higher or lower pitch (frequency) than the last two?
   b. Measure the distance between two consecutive dark crossing spots. __________ cm
   c. Are the dark crossing spots closer together or farther apart than those in Step 2?
   d. What can you say about the distance between interference points as the frequency of a sound decreases?
4. Now overlap #1 and #4. How does the pattern made by the dark and light spots compare to the last two?

Part 2

1. Open the NCH Tone Generator program on your computer.
2. Set up your speakers so that they are not both on the same side as you.
3. Set the program to be in stereo by clicking Menu, Tone, and Stereo (see Figure 51.1, p. 217).
4. Set both the left and right speakers to play a constant sine wave with a frequency of 2,000 Hz (a fairly high frequency) by clicking Tone and Sine and then double-clicking both the Sine 1 Left Frequency and Sine 1 Right Frequency and changing the values to 2,000 Hz.
5. Click the Play button and adjust the volume of the speakers to a comfortable but easy-to-hear level.
6. Plug one ear with your finger and move your head back and forth. You should be able to hear the sound getting louder and softer.
7. Estimate the distance between two loud or two soft spots. __________ cm
8. Now lower the frequency. According to your results in Part 1, do you think that the loud spots are going to get closer together or farther apart?
9. Now change the frequency to 1,000 Hz and find the loud and soft spots again.
10. Estimate the distance between loud or soft spots. __________ cm
11. Were they closer or farther apart than those in Step 7?
12. Predict how far apart they will be when you change the frequency to 500 Hz (a fairly low frequency).
13. Perform the experiment and estimate the distance between the loud spots. _________ cm

Figure 51.1

The Tone Generator Window

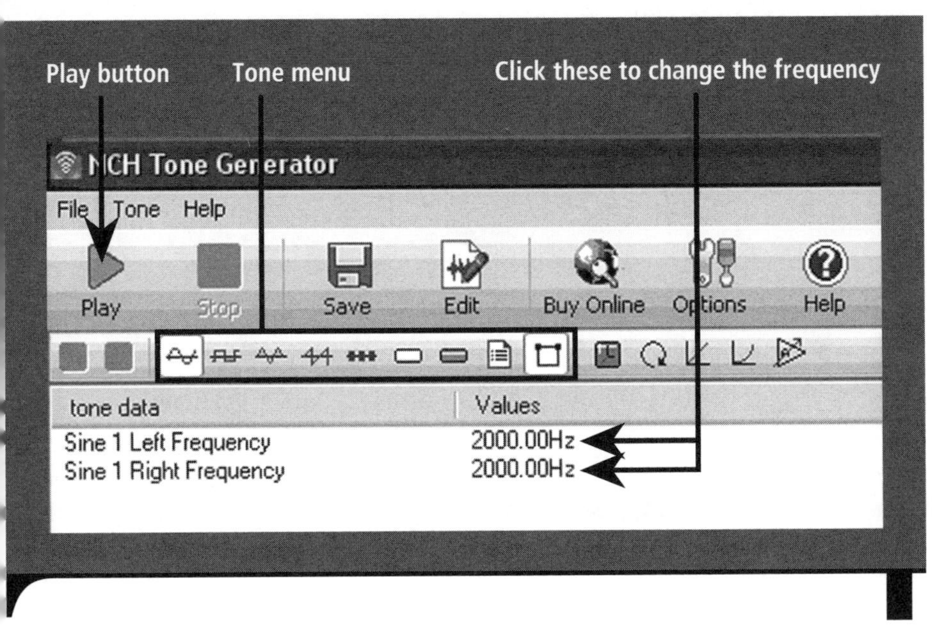

Post-Lab Questions

1. Finish the sentence: As frequency gets higher, the loud spots get ____________________.
2. Sketch a graph of your results with frequency on the *x* axis and distance between loud spots on the *y* axis. Make it so that your graph is as large as a page and it goes from 0 Hz to 20,000 Hz.
3. Use your graph to figure out how far apart the loud spots would be at the extremes of human hearing (20 Hz and 20,000 Hz). Could you experience both of these while seated in your chair?
4. If the thickness of the black line represents loudness, which set of waves was louder? Did this have any effect on the distance between loud spots?

SECTION 4:
Electricity and Magnetism

LAB 52: CREATING STATIC ELECTRICITY

In this inquiry activity, students will get some experience with static electricity before getting deep into electricity and magnetism. They will not know which combinations will generate static electricity and will test some materials of their own choosing.

The lab is intended as an introduction to the unit on electricity. Students should have the basic idea that charged objects may attract or repel each other. They will discover that conductors of electricity cannot be charged with static electricity via friction.

Topic: Static Electricity
Go to: *www.sciLINKS.org*
Code: THP32

Post-Lab Answers

1. Yes, some materials will charge one item and not another. For example, hair will charge a balloon, but will not charge wood.
2. Yes, some items are conductors and cannot be charged by friction (e.g., aluminum foil). Other items are such poor conductors that they will not carry a charge at all (e.g., wood).
3. Yes, balloon and fur, PVC and turkey roasting bag, and silk and glass will all charge well. Other combinations (acrylic and wool) may work well, too.

LAB 52: CREATING STATIC ELECTRICITY

QUESTION

Which combination of materials, when rubbed together, will create static electricity?

SAFETY

Be careful with the glass and wood in order to not get cuts or splinters.

MATERIALS

PVC pipe, glass, balloon, aluminum can, wood, plastic ruler, turkey roasting bag (TRB), paper, wool, fur (your own hair will work if you don't have a lot of hair-care products in it), aluminum foil, silk (other material if you don't have silk—nylon, cotton, rayon)

PROCEDURE

When you rub a material that easily loses electrons with a material that collects extra electrons, the objects get charged with static electricity. Some people believe that only the object with the extra electrons is charged, but both objects are. If you come near the object with extra electrons, you have fewer electrons and hence the electrons will try to flow in a direction that will even out the electrons. The extra electrons in the other material will travel to you. If you come close to the material that lost the electrons, you will look like you have extra electrons. The extra electrons in your body will rush out of you into the other material. Either way, you will feel a small shock. When you feel a shock, you can't tell if it is electrons rushing in or electrons rushing out.

In this activity, you will be rubbing a variety of objects together to see if they get charged with static electricity. You will test for static electricity by seeing if the object will attract small scraps of paper.

1. Collect as many of the materials as possible. Feel free to substitute if you don't have one of the materials. Change the data chart if you substitute a material.
2. Cut up small pieces of paper (<0.5 cm) and put them on a smooth table.
3. Rub each one of the materials listed on the top of the data chart on each of the materials listed on the side of the chart and record whether the material on the left side of the chart attracts the small pieces of paper or not. With some of the combinations (like PVC pipe and TRB), you can hold one material in one hand and whack it with the other material. Rinse the charged object with water to remove any charge before using it again. Find some other materials around the house to test and include them in the chart under "other."

Data Chart

Material	Paper	Wool	Fur/Hair	Aluminum foil	TRB	Other
Plastic ruler	Yes/No	Yes/No	Yes/No	Yes/No	Yes/No	Yes/No
Glass	Yes/No	Yes/No	Yes/No	Yes/No	Yes/No	Yes/No
Aluminum can	Yes/No	Yes/No	Yes/No	Yes/No	Yes/No	Yes/No
Wood	Yes/No	Yes/No	Yes/No	Yes/No	Yes/No	Yes/No
PVC pipe	Yes/No	Yes/No	Yes/No	Yes/No	Yes/No	Yes/No
Balloon	Yes/No	Yes/No	Yes/No	Yes/No	Yes/No	Yes/No
Other	Yes/No	Yes/No	Yes/No	Yes/No	Yes/No	Yes/No

Post-Lab Questions

1. Were there substances that were charged by some materials and not by others? Which were they?
2. Were there substances that were not charged by anything? Which were they?
3. Did some combinations seem to work well? Which were they?

Extensions

How does a Van de Graaff generator make static electricity? Go to *www.scitoys.com* to find out how to make your own miniature Van de Graaff generator. How does a Tesla coil work? How does a Wimshurst generator work?

LAB 53: ATTRACTION AND REPULSION

In this lab, students will experience the interaction of charged objects. Students should realize that the phrase "opposites attract" isn't exactly correct. It is not just positively and negatively charged objects that are attracted. Positively charged objects are also attracted to neutral objects, and negatively charged objects are attracted to neutral objects, too. The phrase would be better stated as "Likes repel, unlikes attract." Water will be attracted to a charged balloon and water itself is not charged.

Post-Lab Answers

1. Answers will vary based on the weather and how vigorously the balloon is charged. Dirt usually does not move unless there are organic particles in it. Sometimes crisped rice cereal moves and sometimes it doesn't. Small pieces of paper and aluminum foil tend to stick. Salt and pepper tend to attract, touch, and repel the balloon.
2. Things that have the same charge repel each other. Things that have unlike charges (+/-, +/neutral, -/neutral) attract each other. This activity would only be good for determining if two objects had the same charge by seeing if they repel each other. It could not discern whether another object is attracted because both neutrally or oppositely charged objects will be attracted. Assuming that all balloons have the same charge (a very safe assumption), then it would work for balloons, but not other objects.
3. It is not just gravity that makes it come to the balloon at first. It is actually attracted. It is uncharged and attracted to the charged balloon. Once it touches the balloon, they have the same charge and repel each other.

LAB 53: ATTRACTION AND REPULSION

QUESTION

How do different charged and uncharged objects interact with each other?

SAFETY

Balloons are a choking hazard for young children.

MATERIALS

2 balloons, salt, pepper, aluminum foil, dirt, crisped rice cereal, small spool of thread, Christmas tree icicle (or tinsel)

PROCEDURE

When you rub a material that easily loses electrons with a material that collects extra electrons, the objects get charged with static electricity. Some people believe that only the object with the extra electrons is charged, but they both are; they both can shock you. If you come near the object with extra electrons, you have fewer electrons and hence the electrons will flow toward you to even out the charges. The extra electrons in the other material will shock you. If you come close to the material that lost the electrons, you will have extra electrons in comparison. The extra electrons in your body will rush out of you into the other material, and you will feel a shock. When you feel a shock, you can't tell if it is electrons rushing in or electrons rushing out.

When you rub a balloon and a furry object together, the furry object loses electrons and the balloon gains electrons. Then if another object touches the balloon, it will take some of the electrons and have an excess of electrons itself. Then they both will have excess electrons. In this lab, you will be charging a balloon with static electricity and using it to perform several experiments.

Step 1

1. Charge the balloon with static electricity by rubbing it on your hair, on a dog or cat, or on a sweater. The faster you rub, the better. Also, this experiment works best if the weather is cold and dry.
2. Put some salt, pepper, small pieces of aluminum foil, dirt, and small cereal (e.g., crisped rice cereal) in separate saucers. Bring the charged balloon near each of them and record your observations.
3. Recharge the balloon after each one.

Step 2

1. Tie two balloons to the end of strings about 2 ft. long. Charge one balloon by rubbing it. Leave another balloon uncharged.
2. Hang the balloons near each other and record what happens.
3. Now charge both balloons and repeat the previous step. Record your observations.

Step 3

1. Charge a balloon with static electricity and turn on a slow stream of water from a faucet in a deep sink.
2. Hold the charged side of the balloon near the stream of water and record your observations.

Step 4

1. Get a piece of Christmas tree icicle (or tinsel), break it in half (~15 cm), and tie the ends together to make a loop approximately 3 in. in diameter. Cut off any excess after tying the ends.
2. Charge one side of the balloon and remember which side is charged. Hold the balloon at arm's length with the charged side facing up.
3. Hold the loop near the balloon until it is attracted to it and let it go. Just as the loop touches the balloon, pull it away, forcing the loop to come off the balloon.
4. With practice, you can get the loop to hover in thin air above the balloon for long periods of time. Don't let it get too close to your body or it will be attracted to your clothing. If the loop sticks to the balloon, pull it off and recharge the balloon.
5. Practice moving the tinsel up, down, left, right, forward, and backward. Just don't get it too close to walls or your clothes or it will stick to them.
6. Generate your own data chart for recording your observations.

Post-Lab Questions

1. Were any of the substances in Step 1 not attracted to the balloon? Did any of them stick and not come off? Is there something special about these materials that would give them this unusual behavior?
2. Explain the behavior of the charged and uncharged balloon and the two charged balloons in Step 2. Could this behavior be used to test if a balloon is charged or not?

3. Explain why the tinsel was first attracted to and then repelled from the balloon in Step 4.

Extension

Try the flying tinsel activity with other charged materials. Try with PVC pipe and roasting bags, ruler and paper, or vinyl record and wool. Would it matter if the material had extra electrons or a lack of electrons for this to work?

LAB 54: SPARK LENGTH

This inquiry activity is performed before students understand the difference between current and voltage. Students also determine the procedure themselves and design the data chart and graph on their own. This makes it Level 3 inquiry.

The lab introduces the foundation that students will need to understand the difference between voltage and current. The sparker puts out very high voltage at a very low current. Current is what is harmful to people, so even though the spark is thousands of volts, it cannot kill a person; however, the 110 volts in a home receptacle will easily kill a person. You can find on the internet the current that it takes for a person to feel it, to be hurt by it, or to be killed by it, and use that to estimate the current put out by the sparker. Students should also be reminded that water is a poor conductor of electricity, contrary to popular belief, in order to understand how humidity affects spark length.

To get the piezo sparker, simply open the lighter and follow the wires leading to the sparker. There will be a wire coming from one end and a metallic base on the other. These are the positive and negative poles. There will also be a button that can be depressed. It will look something like Figure 54.1, depending on the lighter.

Figure 54.1

Piezo Sparker

Post-Lab Answers

1. Inverse—the higher the humidity, the shorter the spark length.
2. The height of the clouds above the ground, the humidity of the air, the temperature of the air.
3. Spark plugs (not piezo), gas stove lighter (some piezo, some not), Bunsen burner lighter (some piezo, some flint), barbecue lighter, candle lighter.

LAB 54: SPARK LENGTH

QUESTION

What is the voltage of a piezo lighter?

SAFETY

Standard safety precautions apply. Do not use the piezo sparker on anything except the resistors.

MATERIALS

Piezo sparker from a lighter, ruler, T-pin

PROCEDURE

Piezoelectric crystals have a crystalline structure that, when disturbed, generate a potential difference. In other words, applying a force to a piezoelectric crystal can cause it to create a spark under the right conditions. This is how most handheld barbecue or fireplace lighters with triggers work. A spring-loaded device strikes the crystal when the trigger is pulled and the resulting spark lights the butane gas within the lighter. The voltage of these devices can be very high. You will estimate the voltage by measuring the maximum length of the spark. The higher the voltage, the longer the spark. Electricity must have a path to the ground to follow or it won't go anywhere, so you will use a metal faucet that goes into the ground to complete your circuit.

Part 1

Given a little bit of background information, you will determine the procedure yourself. You will get the maximum spark from your device if you are sparking it to a pointed object. A T-pin taped to a metal water faucet will be perfect.

Decide what data to collect to determine the voltage of the device. Write down the steps of your procedure. Use illustrations if necessary. Repeat the procedure at least three times. Record your data in a chart that you create. Show any calculations used (for example, 9,000 V will spark across a 1.0 cm gap).

Part 2

1. Use the internet to find the approximate current humidity in your area each day for a week.
2. Use the procedure that you developed to measure the spark gap every day for a week.
3. Draw a graph of humidity versus gap length.
4. Draw your own data charts and graphs as you feel appropriate.

Post-Lab Questions

1. Is the relationship between humidity and spark length direct or inverse?
2. What additional information would you need to be able to estimate the voltage of a lightning bolt from the factors you have studied here?
3. Name three devices that are used in your everyday life that depend on electrical sparks. Do any of them use piezoelectric sparkers?

LAB 55: STATIC SWING

This activity is inquiry-based in that students will perform the activity before they formally learn about how objects can become charged with static electricity (induction, contact, etc.). This will allow the teacher to refer back to this activity when teaching the concept in class. Students will have the same background knowledge and experiences and will have an activity on which to "hang" their knowledge about charging objects with static electricity. Students also should have a basic idea of "complete circuits" from middle school. Students will be guided to understanding through a series of questions.

It is becoming more difficult to find a television with a tube in it. Many home televisions are now flat-screen LCD or plasma, which will not work for this experiment. A tube-style television or computer monitor is needed.

Post-Lab Answers

1. The circuit was not complete until a large or grounded object came in contact with the second wire.
2. It was first attracted because the can became charged and the ball was uncharged. Once it touched the can, then they were both charged with the same polarity so they repelled each other.
3. The ball of foil was charged after touching the first can. Because the second can is grounded, it is uncharged; hence the ball is attracted to it.
4. This device works by carrying charge from one can to another. When the first can becomes charged, it attracts the ball to it. Once the ball touches it, the ball becomes charged and repels away. This causes the ball to be attracted to the second can. When it touches that can, it loses its charge and becomes attracted to the first can again. This process repeats.

LAB 55: STATIC SWING

QUESTION

How can simple materials be used to detect static electricity?

SAFETY

Be careful not to touch any electrical components in the television or monitor. Do not scratch the tube with the aluminum foil.

MATERIALS

2 empty soda cans, thread, aluminum foil, two 12 in. wires with stripped ends, tape (Scotch or other), straw, tube television (not a flat-screen)

PROCEDURE

When an object is charged with static electricity, it is either because it has an excess of electrons or a deficit of electrons. An object with excess electrons and an object with a deficit of electrons will be attracted to each other and may spark when they come close enough to transfer electrons. This works because of the potential difference between them. But there is also a potential difference between a charged and an uncharged object. If an object is connected to the ground or a very large, conducting object, it is considered to be uncharged (also described as being neutrally charged, or having no net charge).

In this experiment, you will be collecting static electricity from a television or computer monitor. This electricity will be used to make an object move. Your task will be to explain how the device works.

1. Collect the following items: two empty soda cans with the pop tabs still intact, a piece of thread, two 12 in. wires, tape, aluminum foil, and a straw or plastic pen.

2. Lift up the tabs on the cans and put the straw through them so that it goes from one to the other (see Figure 55.1, p. 236). Separate the cans by approximately 2 in.
3. Tie the thread to the middle of the straw between the cans and cut it about an inch from the bottom of the cans.
4. Take a piece of aluminum foil approximately 2 in. by 2 in. and wad it up in a ball on the end of the thread. It should dangle between the cans about an inch from the bottom.
5. Cut a piece of aluminum foil the same width as your television or monitor. You are going to be connecting a wire to your foil, but some foil is coated on one side so the wire will have to touch both sides to make contact. Poke a hole near one corner of the foil, strip the end of the wire, and twist it through the hole. Put some tape to hold it on both sides of the foil.
6. Tape the other end of the wire to the side of one of the cans. The can is painted, so you must scrape off some of the paint with sandpaper or the back end of a fork or spoon.
7. Scrape some paint off the side of the other can, too. Strip the ends of the other wire. Connect one end of the wire to the stripped side of the can with tape and leave the other one dangling.
8. Put the two cans, the straw, and the ball of foil on top of the television or computer monitor. Turn on the television or monitor and leave it for a minute or two.
9. Hold the sheet of foil up to the television or monitor screen. Does anything happen to the ball of foil?
10. Turn the television or monitor off and back on. Does anything happen to the ball of foil?
11. Hold the end of the wire that is dangling in one hand; hold the foil up to the television or monitor with the other hand and have someone turn it on. Does anything happen to the ball of foil?

    If nothing happens in Steps 9–11, move the cans a little closer together and try again. If it still doesn't work, take off your shoes.

Figure 55.1

Static Swing Sitting on a Television

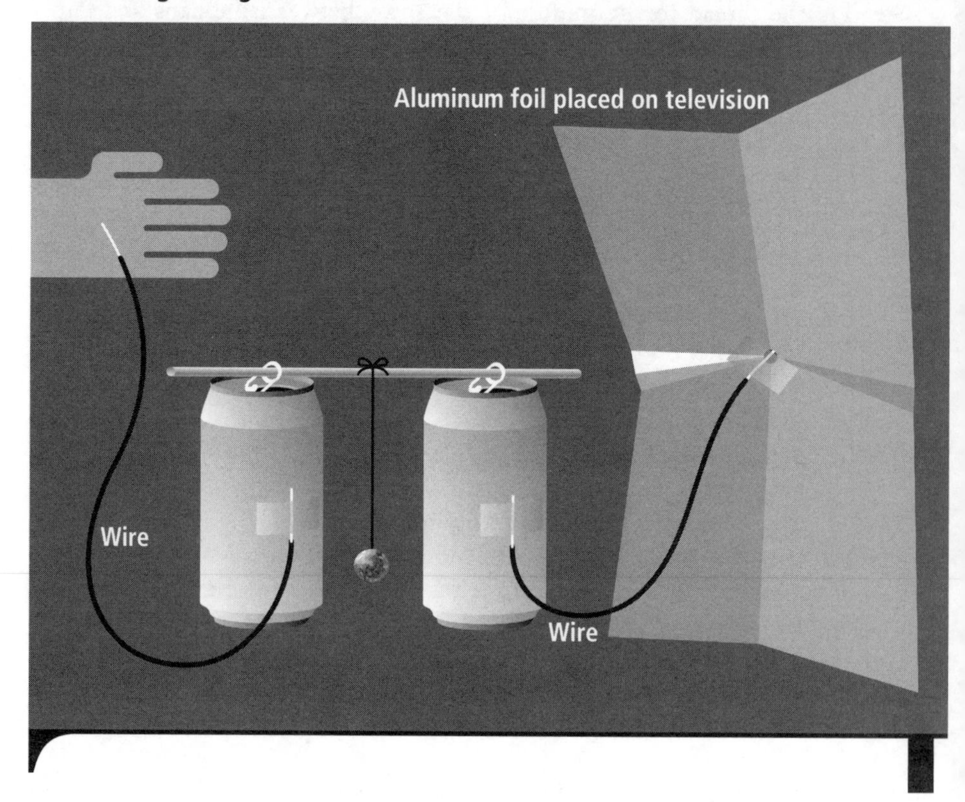

Post-Lab Questions

1. Why didn't the device do much until you touched the dangling wire?
2. What made the ball of foil attract to the first can? Why did it get repelled from the can?
3. Why did the ball of foil attract to the second can? Why did it release from the can?
4. In a paragraph, explain how this device works.

Extension

Benjamin Franklin first conceived a similar type of device in the mid-1700s. He built one to detect when a lightning storm was approaching. He used bells instead of cans so that he could hear it. Search the internet for "Franklin's bells" or "lightning bells" to find out more.

LAB 56: ELECTRICITY AND SAFETY

Discuss with students a common misconception about lightning rods: They do not necessarily attract lightning strikes. They leak charges out of the clouds into the ground so that lightning strikes are less likely to occur. They do occasionally get struck by lightning, but that is not their intended purpose. Be clear that what students are seeing in this activity are charges leaking out of clouds, not a lightning strike.

Students will color diagrams with a graphite pencil, and the thin layer of graphite will conduct electricity. When a barbecue lighter is sparked from the clouds to the ground, electricity will flow along the surface of the layer of graphite and students can see it in a darkened room. Seeing where the electricity flows will help them understand why some activities are safe and some are not safe around electricity.

Post-Lab Answers

Part 1

1. The electricity will pass through the lightning rod. This is because the rod is a good conductor and because it is tall. The charges will either leak through the rod and lightning will be prevented, or the rod may be struck by the lightning and protect the structures around it.
2. The resistance of 100 ft. of metal rod is lower than 100 ft. of brick, wood, or air, so the electricity passes more easily through the metal rod. It follows the path of least resistance.
3. The rod would have the electricity pass through it because it is less resistive than a tree. Trees do get struck by lightning, but not if metal poles are around.
4. a. Not safe. Trees have been known to explode, burn, and fall down when struck by lightning. You do not want to be hugging one if it gets struck.

   b. Not safe. If you are the tallest thing around, you will be the path of least resistance because your body conducts electricity better than air.

c. Relatively safe compared to the other options. Crouching or laying on the ground are good ways to avoid being struck by lightning in many situations.
d. Not safe. You do not want to be holding a metal object high in the air during an electrical storm. Golfing is one of the most dangerous things to do in a lightning storm.
e. Not safe. A wet kite string will likely conduct electricity better than air, so you do not want to hold on to the end of that string. Note that Benjamin Franklin's kite did not have a key near the top that got struck by lightning. It had a key at the bottom that slowly leaked sparks onto a metal object nearby.

Part 2

1. Yes, the electricity still passed through the car. A centimeter of rubber is not enough to stop lightning from striking. Tires do not protect a car from being struck by lightning.
2. No, the lightning did not pass through the person. It is easier for it to pass through the metal body of the car than to spark through the air, travel through the person, and then jump back to the car frame. It follows the path of least resistance, which is the metal parts of the car.

Part 3

1. No, the electricity did not pass through the bird. It is easier for the electricity to continue through the couple of inches of wire between the bird's legs than to jump through the wire insulation, pass through the bird, then jump back through the insulation again. Electricity going through the bird would encounter much higher resistance than simply going through the wire. It followed the path of least resistance.
2. Yes, the electricity passed through the bird the second time. If there is a different voltage on the two wires, then there is a potential difference between them and electrical current may flow. If the bird touches a power line and a telephone line, a power line and a metal tower, or a power line and any other conductor connected to the ground, it will be shocked. It is not possible for the average person to tell if a cable carries electricity or not, so do not ever touch a power line.
3. Anything that has power running through it is different than zero, so there is a potential difference. If there is a potential difference, current may flow through you.

Note to teachers: Reproduce the following page separately for students.

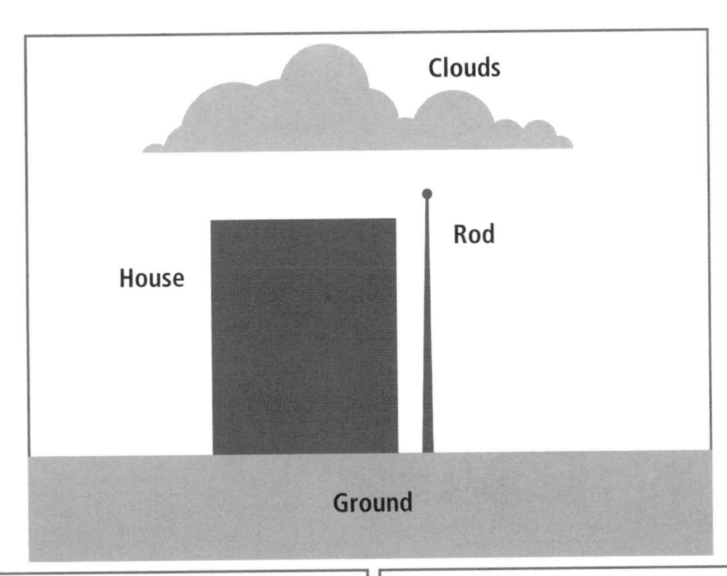
Clouds
Rod
House
Ground

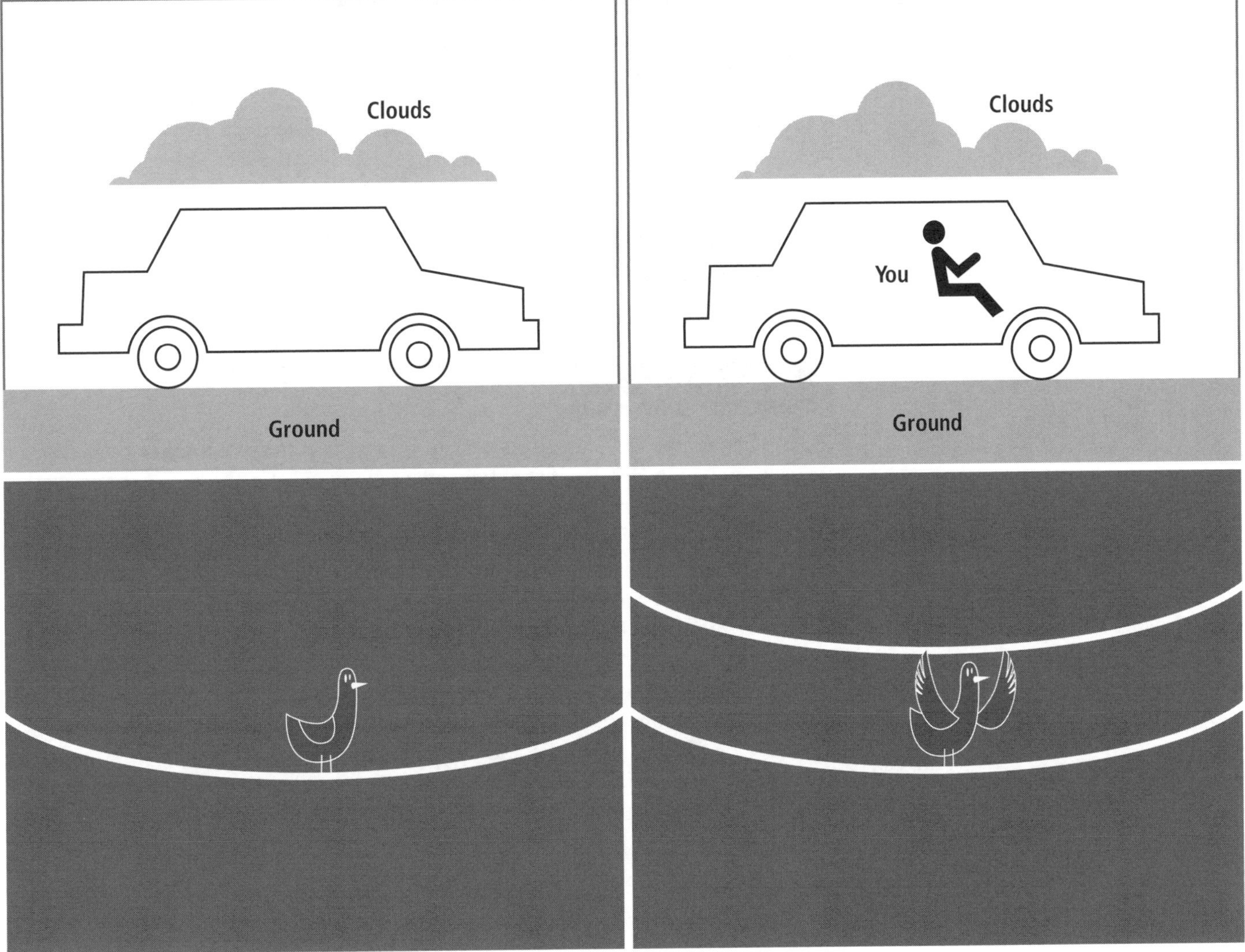
Clouds
Ground
Clouds
You
Ground

LAB 56: ELECTRICITY AND SAFETY

QUESTION

Why do people put lightning rods near tall buildings? Why is it safe to be in a car in a lightning storm? Why does a bird not get electrocuted when sitting on a wire?

SAFETY

Do not experiment with electricity or lightning. These are just simulations and electricity is very unpredictable and dangerous. Never touch power lines. Do not use the piezo sparker on anything except the resistors.

MATERIALS

Graphite pencil, piezo sparker from lighter

PROCEDURE

Electricity can be very dangerous and it is important to understand a little about it in order to be safe. Many of the concepts of electrical safety can be summarized in the statement "Electricity follows the path of least resistance." The two important parts of this concept are that electricity must follow a path or circuit to the ground or back to its source and that if the electricity has the option of two paths, most of it will go through the path that has the least resistance. For example, you don't get shocked when you touch a lightbulb because it is easier for the electricity to pass through the filament and back to the plug than to arc through the gas in the bulb and travel through your entire body to the ground and through the soles of your shoes. But you don't want to change lightbulbs on a metal ladder because you can become the path of least resistance and get shocked.

Graphite conducts electricity. In this lab, you will use graphite from a pencil and the sparker from a barbecue lighter to test some electricity safety scenarios.

You will be able to see the spark traveling along the surface of the graphite in a darkened room to tell where the electricity flows.

Part 1

In places where thunderstorms are common, many people have lightning rods next to their house or on top of the house with a wire leading to the ground. The rod is slightly taller than the house and is connected to the ground.

1. Using your graphite pencil, color very darkly the house, rod, ground, and clouds in the copy of Figure 56.1 that your teacher gives you.
2. Connect the spark generator from the clouds to the ground. If your generator has two wires, touch one to the clouds and one to the ground. If it has only one wire, it has a metallic base (usually brass) that acts as the other wire. Set the metallic base on the clouds and touch the wire to the ground.
3. Turn out the lights and spark the piezo device (usually by squeezing it until it clicks). Watch for the pathway that the spark follows as it skips across the graphite.

Figure 56.1

Lightning Rod Simulation

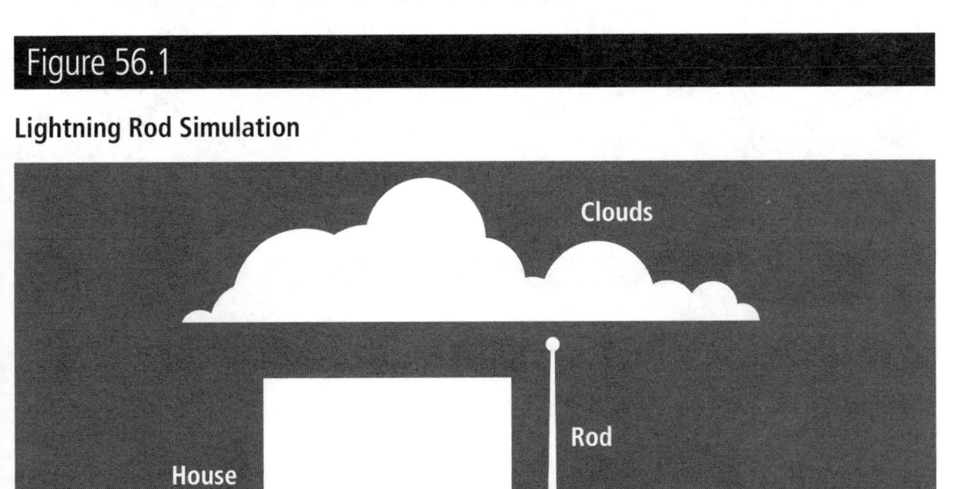

Part 2

In this section, you will be investigating why it is safe to be in a car during a lightning storm even if the car is struck by lightning. There is a common misconception that it is the rubber tires that protect the car, but if that were the case, the car would never be struck in the first place.

1. Color in very darkly the entire car, the rims, the clouds, and the ground on the copy of Figure 56.2 that your teacher gives you. Leave the tires uncolored to represent the rubber tire insulators.
2. Connect the spark generator from the clouds to the ground as you did in part 1, turn out the lights, and spark it. Observe whether or not the electricity flows through the car.

Figure 56.2

Car Being Struck by Lightning

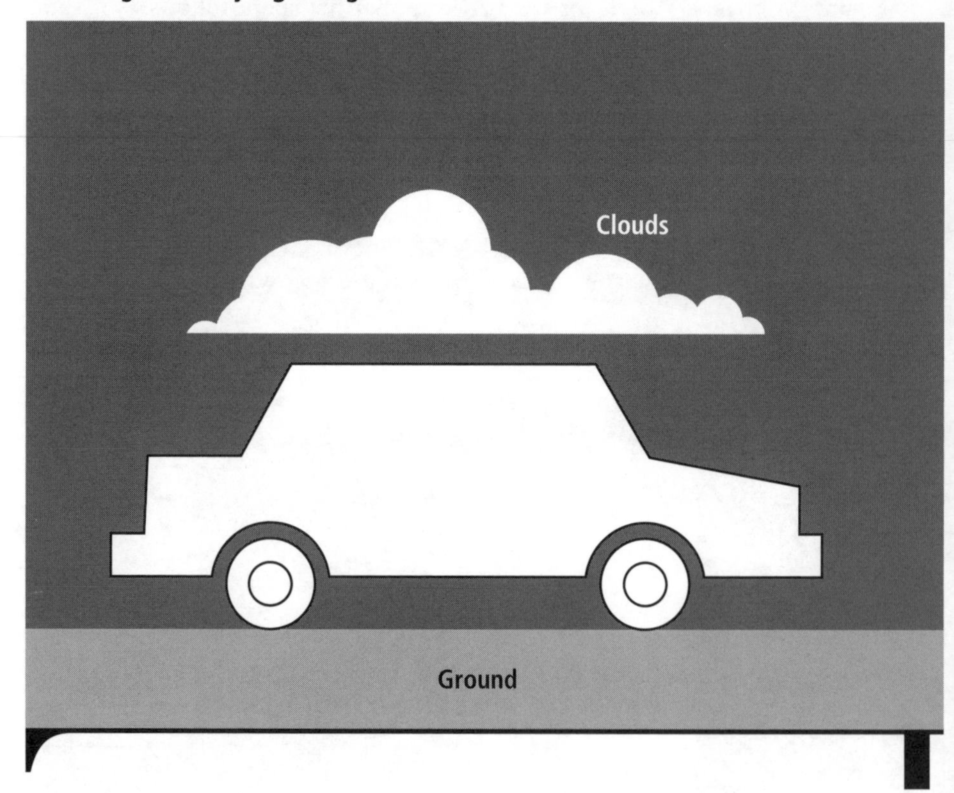

3. Now color in only the electrical conductors (outline of the car, the rims, and the passenger) of the copy of Figure 56.3. Spark from the cloud to the ground and see where it goes.

Figure 56.3

Car (With Passenger) Being Struck by Lightning

Part 3

In this section, you will be investigating why a bird is unharmed when it sits on a single wire, but may be injured if its wings contact two wires.

When drawing resistors, the wider the resistor, the less the resistance. Because a bird has much higher resistance than a few inches of copper wire, in this simulation the wires are drawn with thick lines and the birds will be drawn with thin lines.

1. Color in very darkly the power line and the outline of the bird in your copy of Figure 56.4. Connect the spark generator from one end of the line to the other, dim the lights, and spark it. Observe the path that the electricity follows.

Figure 56.4

Bird Sitting on a Power Line

2. Now color in the two power lines and the outline of the bird very darkly on your copy of Figure 56.5. Connect the spark generator from one end of the first line to the other end of the second line. Dim the lights, spark it, and observe the path.

Figure 56.5

Bird Touching Two Wires

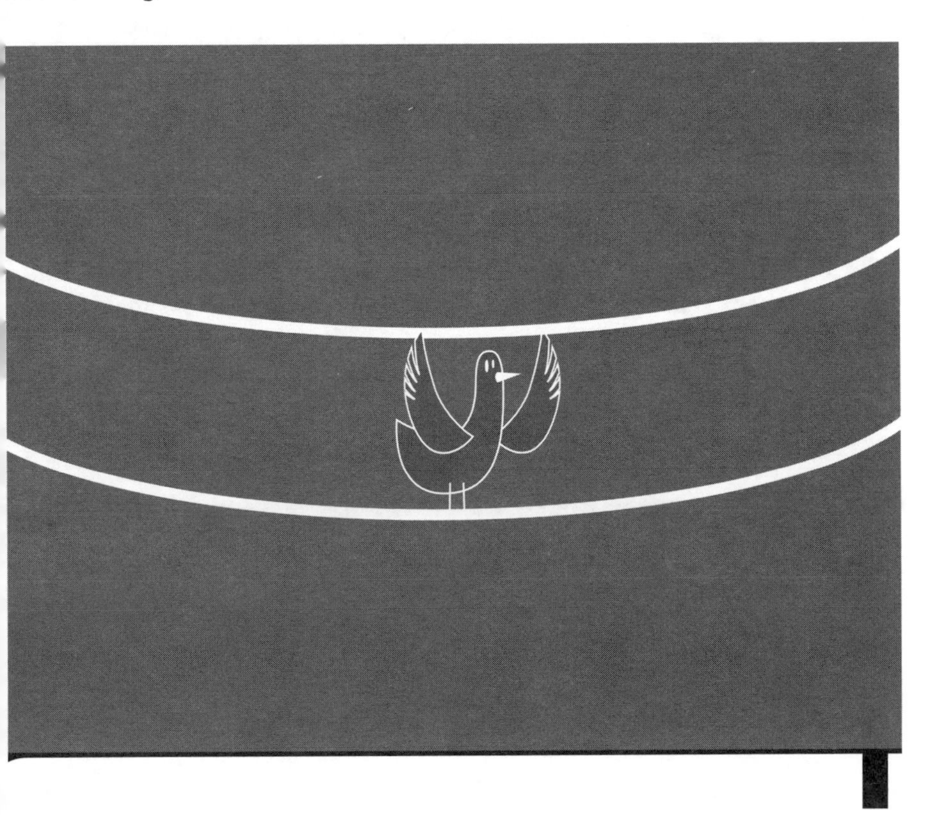

Post-Lab Questions

Part 1

1. What path did the electricity follow? How would this benefit the people in the house?
2. There is a rule of thumb that says that electricity follows "the path of least resistance." How could this be used to explain the operation of a lightning rod? (Hint: Air has a very high resistivity.)
3. If there was a tree next to a lightning rod and they were the same height, which one would likely have electricity pass through it? Explain your answer.
4. If you were outside during an electrical storm, discuss how safe each of the following activities would be:
   a. Hug a tree
   b. Stand in the middle of a field
   c. Crouch down as low as possible
   d. Go golfing
   e. Fly a kite

Part 2

1. Did the lightning travel through the car the first time even without the tires colored in? What does this tell you about how conductivity can change with very high voltages?
2. Did the lightning travel through the person in the car in the second instance? Use the "path of least resistance" argument to explain your observations.

Part 3

1. Did the electricity pass through the bird in the first trial? Use the "path of least resistance" argument to explain this.
2. Did the electricity pass through the bird in the second trial? Use the idea of "potential difference" to explain this and what conditions must be present for it to happen.
3. Anything connected to the ground is considered to have zero voltage. Why is it dangerous to touch an electrified wire while standing on the ground?

LAB 57: BATTERY AND LIGHTBULB

The number of batteries used in this lab is determined by the type of lightbulbs that are used. A 3 V lightbulb will require two batteries. Students should perform this activity after being instructed how to use the multimeter. Instructions will vary based on the model of multimeter. Students will create a circuit with a battery and lightbulb and use a multimeter to analyze the circuit.

Topic: Batteries
Go to: *www.sciLINKS.org*
Code: THP33

Topic: Light Bulbs
Go to: *www.sciLINKS.org*
Code: THP34

Post-Lab Answers

1. Answers will vary greatly based on the lightbulb used.
2. Answers will vary greatly based on the lightbulb used.
3. Answers will vary greatly based on the lightbulb used.
4. The biggest source of error is that the resistance of the bulb changes as it warms up. The battery and wires also heat up when in operation. The voltage of the battery when it is in use can be different than when it is not in use.

LAB 57: BATTERY AND LIGHTBULB

QUESTION

What are the resistance, voltage, and current of a circuit with a battery and lightbulb?

SAFETY

Do not touch the lightbulb; it can get hot. Handle batteries carefully; they may get hot.

MATERIALS

2 D-cell batteries, lightbulb, 4 6-in. wires stripped at both ends, multimeter

PROCEDURE

In this lab, you will be connecting a small lightbulb to a battery and measuring the resistance, voltage, and current in the circuit. It is important to remember that a voltmeter is connected in parallel to (across) the object being measured and an ammeter has to be inserted into the circuit by breaking the circuit and having it pass through the meter. Be sure that you know how to connect the ammeter or you will damage the meter.

1. Use the multimeter to measure the voltage of the battery. Be sure that the meter is set on DC volts and set around 5 or 10 V. Touch the red probe to the positive terminal (marked on the battery) and the black probe to the negative terminal.
   _________ V
2. Now stack two batteries on top of each other (in series with each other) and measure the voltage again from the top of one battery to the bottom of the other.
   _________ V

3. Measure the resistance of the lightbulb with the multimeter. Set the meter on the highest resistance rating (Ω) and touch one probe to the bottom of the bulb and one to the side of the base. If the meter reads "0" (as it likely will), slowly turn the knob to lower and lower settings until the meter reads a number. __________ ohms
   (Be sure to note whether your answer is in ohms, kilo-ohms, or mega-ohms.)
4. Use Ohm's law ($V = IR$) to calculate the theoretical current through the lightbulb with both batteries connected.
   __________ amps
5. Now connect the bulb to the batteries and measure the voltage of the batteries while the bulb is lit (voltage "under load").
   __________ V
   (If your meter doesn't measure current, stop here.)
6. Your teacher will teach you how to use a multimeter to measure current. Each meter is a little different. No matter which meter you use, you must break the circuit and insert the meter into it. Record the current running through the bulb.
   __________ amps
7. Draw a schematic diagram of the circuit that you built in Step 5 using the correct schematic symbols.

Post-Lab Questions

Note: Use "Voltage under load" (while in operation) for all the voltages here.

1. Use $P = IV$ to calculate the power of the lightbulb.
   __________ W
2. Use $P = I^2R$ to calculate the power of the lightbulb.
   __________ W
3. Use $P = V^2/R$ to calculate the power of the lightbulb.
   __________ W
4. Give two reasons why your answers to the previous three questions may not be identical.

Extension

Go to the website of a battery manufacturer and look up the specifications for a battery similar to the one you're using. Using the data from this lab, figure out how long the battery would be able to power the lightbulb. You will need to find out how many milliamp-hours or watt-hours the battery can provide. If you have a battery with 10 watt-hours, it can power 10 W for 1 hr., 1 W for 10 hrs., or 5 W for 2 hrs.

LAB 58: BATTERY AND LED

An LED should never be connected directly to a battery or it will be damaged. A current-limiting resistor must be attached to either leg to prevent this damage. The value of the resistor will depend on the battery that you use and the LED that you use. Be sure that when you purchase your LEDs, they are rated for a certain current and voltage (preferably in the range of 1.5 V so that a single battery may be used).

Calculate the value of the resistor to be used by determining how much current would flow through that resistor in the absence of the LED. For example, if you are using a 1.5 V battery and an LED rated at 10 mW, you would use Ohm's law to calculate the resistance needed.

$R = V/I$ or $R = 1.5\text{ V}/.010\text{ W} = 150\text{ Ohms}$

In this case, you would attach a 150 Ohm resistor to one of the LED's legs. Soldering is preferred, but twisting the legs together with electrical tape would also be acceptable.

Instruct students that they should not use any other type of battery to power the LED, or they could damage it or injure themselves if the wires overheat.

Students will create a circuit similar to the battery and lightbulb circuit and analyze it with a multimeter. They should see that the LED takes far less energy than a lightbulb.

Post-Lab Answers

1. Answers will vary but may include portable music devices, CD players, DVD players, computers, stereos, laptops, new car taillights, pocket lasers, remote controls, an optical mouse, a ball-type mouse, or toys.
2. Answers will vary, but the LED should have far lower power than the lightbulb.
3. Answers will vary. For example, if the power measured is 20 mW, it would take 50,000 hrs.
4. If a component gets hot, much of its energy is being converted to heat, and it is therefore inefficient. Lightbulbs get very hot; LEDs do not.

LAB 58: BATTERY AND LED

QUESTION

How do you connect an LED to a battery to get it to light up? What are the advantages of LEDs over light bulbs?

SAFETY

Do not connect the LED to more than the recommended voltage or it will get very hot and will be destroyed. Do not connect an LED without a resistor.

MATERIALS

D-cell battery, wires, LED (with current-limiting resistor), multimeter

PROCEDURE

An LED is a light-emitting diode. LEDs are common as little colored lights on computers, televisions, and other appliances. The lights on your computer, computer speakers, keyboard, monitor, and printer that tell you that the power is on are probably all LEDs. LEDs are different than lightbulbs in two main ways. First, they require far less power than a lightbulb does. You would be wasting a lot of electricity if you were to replace every LED in your house with a small lightbulb. Also, LEDs only work when electricity passes through them in the correct direction. A diode is a component that only allows electrons to flow in one direction. Diodes are commonly used for protection in circuits, and in combination with each other can be used to rectify the AC coming out of your wall (which is the first step in converting AC to DC to charge batteries and power small appliances). An LED is simply a diode that gives off light.

An LED normally has one leg (the metal pins sticking out of it) that is longer than the other. The legs will help you determine which way to hook it up in the circuit. Some LEDs also have one side of their body that is flat while the rest of the body is round. LEDs are not meant to handle a lot of current, so a resistor must be connected to one of the legs to limit how much current flows through.

1. Using wires, connect the LED so that the longer leg is connected to the positive pole of the battery and the shorter leg is connected to the negative pole of the battery. Does it light up?
2. Now connect the LED so that the longer leg is connected to the negative pole of the battery and the shorter leg is connected to the positive pole of the battery. Does it light up?
3. Write a rule for connecting LEDs in a circuit.
4. Does the LED feel hot like a lightbulb?
5. Measure the voltage of your battery using the voltmeter.
   _______ V
6. Measure the current running through the LED when it is lit.
   _______ amps
7. Calculate the power running through the LED/battery/resistor circuit.
   _______ W
8. Calculate the resistance of the LED (remember to subtract the resistor).
   _______ ohms
9. Draw a schematic diagram of your circuit using the correct schematic symbols.

Post-Lab Questions

1. Name five things in your house that have LEDs in them.
2. How did the power of the LED compare to the power of the lightbulb in Lab 57: Battery and Lightbulb?
3. If electricity costs 15¢ per kWh, how long would you have to run this LED to cost 15¢? (A kWh is 1,000 W for 1 hr., or 1 W for 1,000 hrs., or 5 W for 200 hrs., etc.)
4. What does the temperature of the LED tell you about its efficiency?

Extension

Find out other ways that LEDs are used in your everyday life. You might be surprised to find out that a lot of inexpensive lasers use LEDs as their source of light.

LAB 59: THE ELECTRICAL SWITCH

This activity is inquiry-based in that students will perform the activity before they formally learn about how a switch works. They will use a multimeter to discover the different settings on a triple-throw switch. The teacher should show the students how to use their multimeters to perform a continuity test.

Switches are very simple and intuitive devices that will be easy for students to understand. In order to complete the circuit described in this activity, students should already know how to hook up an LED, a battery, and a lightbulb.

Topic: Electric Circuits
Go to: *www.sciLINKS.org*
Code: THP35

Post-Lab Answers

1. The triple pole switch works just like its diagram. When flipped to the left, two conductors touch and complete a circuit. In the middle, the conductors do not touch and the circuit is off. When flipped to the right, two different conductors touch and a different circuit is completed.
2. I would have an almost complete circuit on the left with one of the wires broken and an almost complete circuit on the right with the same wire broken. Touching the center wire to the left circuit would turn it on and touching the broken wire to the circuit on the right would make it work.

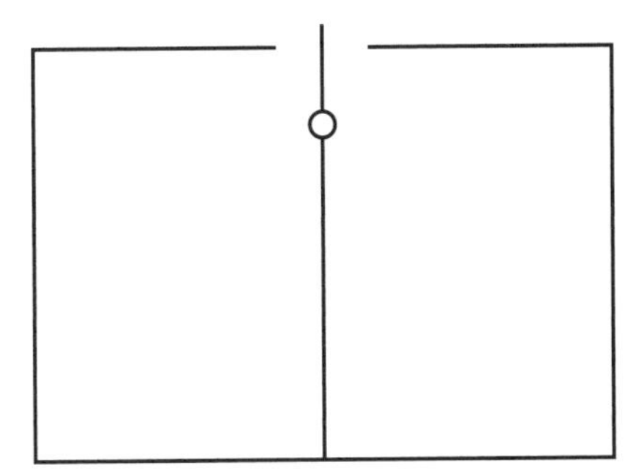

3. Answers will vary. Television, computer, radio—single throw (on/off). Forward/reverse—double throw. Car headlight switch (off, parking lights, headlights)—triple throw. Ceiling fan (off, high, medium, low)—quadruple throw.

LAB 59: THE ELECTRICAL SWITCH

QUESTION

How do electrical switches work?

SAFETY

Never leave a circuit connected too long, as it may get hot and destroy the battery.

MATERIALS

Triple-throw switch (left, center, right), wire, lightbulb, D-cell batteries, LED

PROCEDURE

There are many types of switches in use in modern appliances. There are simple on/off switches that can be in the form of a push button or a light switch. There are low/medium/high switches like the one on a ceiling fan. There are forward, reverse, and stop switches like in a remote control car. Basically, a switch works by making contact between two conductors to complete a circuit. Switches are categorized by how many throws (settings) they have and how many sets of conductors (poles)—for example, SPST (Single Pole Single Throw) or DPDT (Double Pole Double Throw).

In this lab, you will use a switch with three throws, meaning that you can flip the switch to the left, center, or right. If the center position is "off," this is called a Double Throw Center Off switch. In this lab, you will use your multimeter to test your switch. You will use the "continuity" or "diode checking" feature on your meter if it's available. This setting beeps whenever you touch the probes to something that is a complete circuit. If you don't have a continuity setting on your meter, you can use resistance to do the same. When a switch is in the off position, the meter will read infinite resistance or "OL" (overload). When it is in the on position, it will read nearly zero resistance. On your meter, look for either the sound symbol •))) or the diode symbol ▶|.

The schematic symbol for a double-throw switch (left/right) is

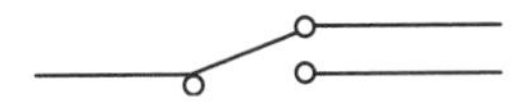

and it looks like Figure 59.1.

Figure 59.1

Double-Throw Switch (Notice the two "legs" on the bottom.)

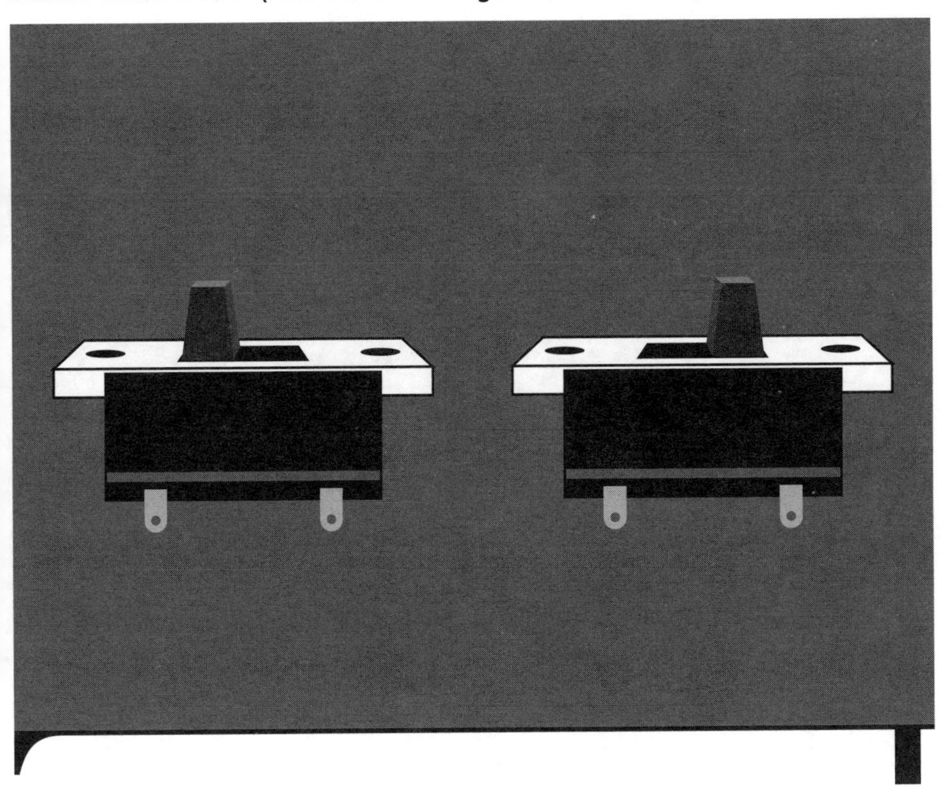

A triple-throw switch (left/middle/right) has the schematic symbol

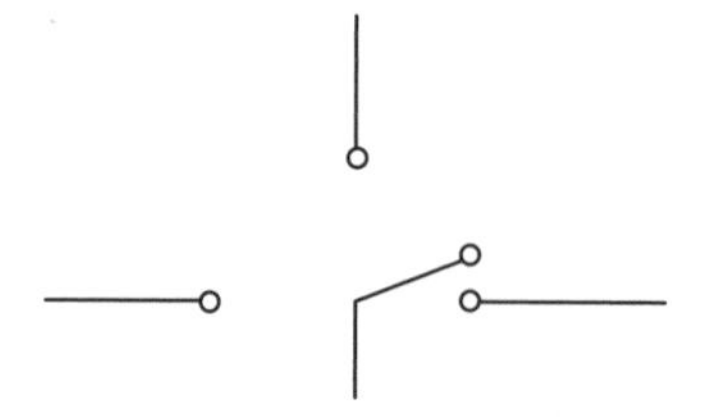

in the open position and it looks like Figure 59.2.

Figure 59.2

Triple-Throw Switch (Notice the three "legs" on the bottom.)

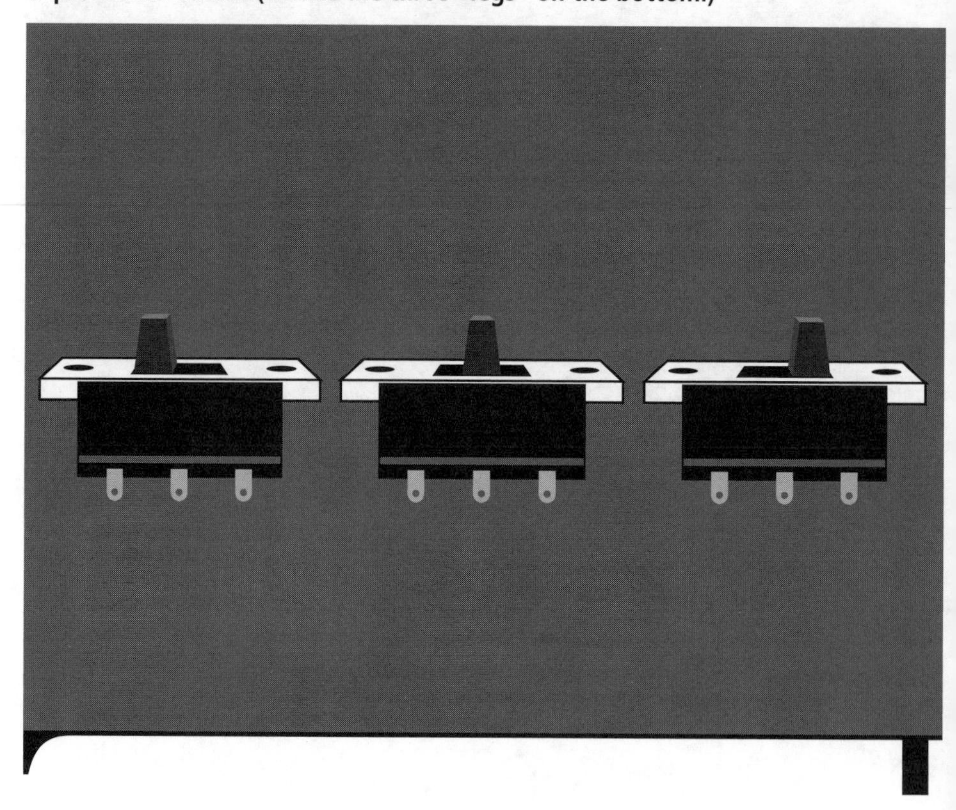

1. Take your switch and mark one side so that you can keep the same side facing you the whole time. Put your meter either on continuity test, diode checker, or resistance and connect the black probe to the left lead and the red probe to the middle lead on the switch. Place the switch in each of the three positions and record whether the switch is continuous or not:

   Left yes/no   Middle yes/no   Right yes/no

2. Now put the probes on the middle and the right leads and repeat Step 1:

   Left yes/no Middle yes/no Right yes/no

3. Now put the probes on the left and right leads and repeat Step 1:

   Left yes/no Middle yes/no Right yes/no

4. Now connect a circuit that will allow you to move the switch to the left and an LED will light up. When you move it to the right, the lightbulb will light up. Draw a schematic diagram of your circuit. (Be sure that your LED has the resistor connected before starting.)

Post-Lab Questions

1. In your own words, explain how a triple-pole switch is used.
2. Explain how you could make a double-pole switch with just wires. Draw a diagram.
3. Find at least five things in your house that have switches in them and tell how many throws they are. Try to find different types of switches.

Extension

Draw a schematic to show how two double-throw switches could be used to control a light from either end of a hallway so that flipping either switch would change the status of the light. Build your circuit with a flashlight bulb and see if it works.

LAB 60: ELECTROMAGNETS

Remind students not to keep the electromagnet connected too long or the battery and wire can get hot. Even the bell wire that appears to be copper colored has enamel on it, and the ends must be stripped to connect to the battery. The wire should be at least 1 m long in order to increase the resistance and decrease the current running through the wire. Be sure that students do not hook the paper clips onto the nail; they should only stick them magnetically to it. Some of the answers may vary depending on the wire, battery, and number of coils used.

Students will construct electromagnets and test to see which variables make them stronger. Although elementary and middle school teachers are expected to have done this activity with their classes, teachers of younger students often do not have access to the equipment. Therefore every student should complete the activity to understand transformers, electric motors, speakers, and generators.

Topic: Electromagnet
Go to: *www.sciLINKS.org*
Code: THP36

Post-Lab Answers

1. a–e: Stronger
   f: Weaker
2. An iron core drastically increases the strength of the electromagnet.
3. The battery voltage increases the strength of the magnet as long as the battery is capable of providing enough current.

LAB 60: ELECTROMAGNETS

QUESTION

What affects the strength of an electromagnet?

SAFETY

Do not keep an electromagnet hooked up for more than 15 seconds or the wire and the battery will get very hot and the battery will die. Do not plug any of these items into a wall socket or connect them to a car battery or other large battery.

MATERIALS

2 D-cell batteries, enameled bell wire, iron nail, small compass, multimeter, sandpaper

PROCEDURE

In this lab, you will make an electromagnet and study how different things affect its strength. First, you will see how an iron core affects your magnet. Build your electromagnet by stripping an inch of the enamel (the colored paint) off each end of your wire with sandpaper. Then wrap the wire around an iron nail. Test your D battery to see that it is not dead. It needs to be more than 1.25 V to be considered good. Be sure to use a long piece of wire or you will short out the battery, and both the battery and wire will get hot while the battery quickly dies.

Part 1

1. Connect the electromagnet to the battery. It doesn't matter which way you connect the wires to the battery. Put one wire on top and one wire on bottom. Don't leave the electromagnet hooked up for more than 15 seconds at a time or it will get hot and your battery will die.

2. Determine how many small paper clips you can pick up with this electromagnet. Open one paper clip and form a hook out of it. Stick it to the electromagnet and then hang more paper clips from the hook. Don't connect the paper clip to the nail. It should be sticking to it, not hanging on it.
3. Now slide the nail out from the coil of wire and see how many paper clips you can pick up with just the coil.

Electromagnet with iron core ________ paper clips

Electromagnet without iron core ________ paper clips

Part 2

Now you will see if voltage makes a difference in the power of a magnet. Test both of your batteries to make sure that they are not dead.

1. Connect the electromagnet to the batteries connected in series (stacked on top of each other. Don't leave it hooked up for more than 15 seconds.
2. Determine how many small paper clips you can pick up with the electromagnet.
3. Slide the nail out and see how many paper clips you can pick up.

Electromagnet with iron core ________ paper clips

Electromagnet without iron core ________ paper clips

Part 3

Now you will determine how the direction of the battery affects the poles of the electromagnet. You will do this by bringing the tip of the electromagnet near a small compass. Whether the tip of the compass is attracted or repelled by the magnet will tell you its polarity. Remember that the Earth's north pole is actually a magnetic south pole. So, the tip of the electromagnet that attracts the tip of the compass is also a magnetic south pole. If it is repelled, then it is a magnetic north pole.

1. Attach your electromagnet with the iron core to the battery in such a way that the wire from the flat end of the nail is connected to the positive side of the battery and the wire from the pointed end of the nail is connected to the negative terminal.
2. Use the compass to determine if the pointed end of the nail is north or south.
3. Now switch the terminals of the power supply and repeat the test. Is the pointed end now north or south?

BE SURE TO SAVE YOUR WIRE WHEN YOU'RE DONE. YOU MAY USE IT AGAIN.

Post-Lab Questions

1. Give your best guess of how each of the following things would affect the strength of the electromagnet by putting "stronger" or "weaker." If you don't know the first four, design an activity to figure them out.

   a. more coils of wire ____________

   b. using a wire with a lower resistance ____________

   c. a thicker nail ____________

   d. using a thicker wire ____________

   e. using a car battery ____________

   f. using AC instead of DC ____________

2. How did an iron core affect the strength of the magnet?
3. How did the voltage of the battery affect the strength of the magnet?

Extension

Find out if electromagnets are really used to pick up cars in junkyards or if that is only in the movies. If they are used, find out more about them, such as how strong they are and how much current flows through them.

LAB 61: MAGNETIC FIELD LINES

Magnets can be purchased inexpensively at discount stores. Sets of refrigerator magnets of letters or numbers can be purchased for less than a dollar, and they usually contain 50 or more magnets. Each item contains two or more magnets that can be popped out with a screwdriver. Note that some of these magnets have their poles at the ends and some have them on the top and bottom. Figure out which type you have and mark them accordingly.

Mark each of the magnets with some paint to identify common poles. It really doesn't matter which pole is which color as long as all of the north poles are one color and south poles are another. Just choose a magnet and paint half of it. Then find the side of the other magnets that repel the painted end and paint them, too.

This activity will introduce students to the idea of magnetic field lines—a concept they have probably encountered but may not fully grasp. Completing this activity and reading the corresponding background information should enable students to understand classroom discussions about magnetic fields and field lines.

Topic: Magnetic Fields
Go to: *www.sciLINKS.org*
Code: THP37

Post-Lab Answers

1. Answers will vary but should be consistent with the student's drawing.
2. This answer will likely come out wrong because the north-seeking end of the compass needle is the south pole of the magnet. The south pole of the compass needle will be attracted to the north pole of the magnet, not repelled. Students should understand that when using a compass to draw magnetic field lines, they should draw the arrows opposite the compass needle. Taking them through this realization will help them remember this fact.
3. Answers will vary. Whichever end it is attracted to is the north pole of the magnet. This will not match with the answer to #2 for the reasons discussed above. The answer to #3 is the correct answer. All of the magnetic field lines in #2 are backward.

LAB 61: MAGNETIC FIELD LINES

QUESTION

How are magnetic field lines drawn?

SAFETY

Small magnets can present a choking hazard for young children.

MATERIALS

2 small magnets, compass, paper

PROCEDURE

Magnetic field lines are a model for thinking about how magnets act. Although magnetic field lines do not really exist, they are a convenient way to model magnetic fields. The magnetic field lines represent the direction that magnetic fields are exiting and entering a magnet and how relatively strong the magnetic field is at a certain point. Keep in mind that there *is* a magnetic field between the field lines.

Note that the magnetic field lines that you see in books follow along a line where the magnetic strength is the same. Because we cannot measure the magnetic strength, we will have a model of magnetic field lines that will not be exact.

1. Place a small magnet under a sheet of paper. One end has been marked with a color; note that in your drawing. Use a pencil to draw the outline of the magnet.
2. Place your compass on the paper near the magnet. Notice which way the arrow (north-seeking end) of the compass needle is pointing. If it is pointing north, make sure that it is close enough to the magnet.

3. Lift the compass and draw an arrow on the paper in the same direction that the north-seeking end of the compass needle was pointing. Make a note of which direction north is. If the compass points that way, it could be too far from the magnet.
4. Repeat this procedure as you move the compass around the magnet (see Figure 61.1). Do at least 20 measurements at different distances and on different sides.
5. Magnetic field lines are continuous coming out of one side of the magnet, going all the way around and then entering the other side. Draw circles that go through each one of your arrows and then symmetrically back to the other end of the magnet. This approximates a magnetic field line.

Figure 61.1

Model of the Magnetic Field Lines Setup

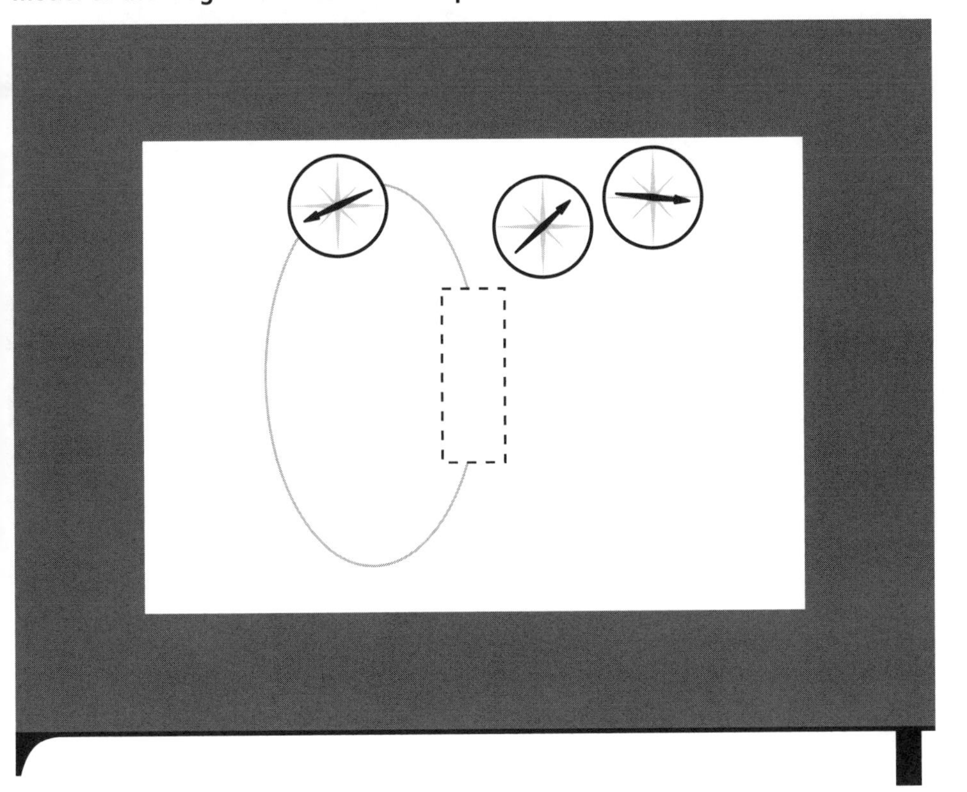

Post-Lab Questions

1. Where did the magnetic field lines come out of the magnet (compass pointed away from the end of the magnet)?
2. By convention, magnetic field lines come out of the north pole and enter the south pole of the magnet. Which end of the magnet is the north pole according to the magnetic field lines?
3. Was there actually a magnetic field between the lines that you drew? Does the space between magnetic field lines mean that there is no magnetic field there? Put the compass between two lines to see.

Extension

Perform this experiment again, but with the two magnets in different configurations. Separate them by a few centimeters with like poles facing each other and map the magnetic field lines between them and around them. Repeat with opposite poles facing each other.

LAB 62: RESISTIVITY EQUATION

Students must understand what direct and inverse relationships are and how the equations for a direct or inverse relationship are written. There are many other ways to determine resistivity, but knowing the diameter of the lead refills makes this method very convenient.

Having students create their own data charts will increase the level of scientific inquiry. Do not be tempted to create the chart for them.

Post-Lab Answers

In the procedure, students should determine that the equation is $R = \rho L/A$ because the larger diameter lead has lower resistance and the resistance of the lead increases as the length increases.

1. Yes, if the shape is such that the cross-sectional area may be calculated, the resistance can be measured and then the resistivity calculated. The object must be regularly shaped with a constant cross section (cylinder, cube, rectangular solid, not pyramid, sphere, irregular).
2. It has lower resistance and will generate less heat.
3. The resistance will be high and much of the power will be wasted.

LAB 62: RESISTIVITY EQUATION

QUESTION

What is the resistivity of graphite?

SAFETY

Standard safety precautions apply.

MATERIALS

0.5 mm pencil lead, 0.7 mm pencil lead, multimeter, ruler

PROCEDURE

Resistance is how much an object resists the flow of electrons. This depends on the material (its resistivity) and the size and shape of the sample being tested. In this lab, you will see whether resistance is directly or indirectly related to cross-sectional area and length and then calculate the resistivity of graphite.

It is known that resistance depends on all three variables: resistivity, cross-sectional area, and length. Resistivity is measured in such a way that it is directly proportional to resistance. If two different materials are exactly the same shape, the one with the higher resistivity will have the higher resistance. Therefore, the equation for resistance can only take on one of the following forms: $R = \rho LA$, $R = \rho A/L$, or $R = \rho L/A$ where R = resistance, ρ = resistivity, L = length, and A = cross-sectional area.

You will be using pencil "lead" (which is actually made of graphite, not the metal lead) to determine the resistivity of graphite.

1. Find a piece of 0.5 mm pencil lead (the thinner of the two pieces). Set your multimeter on the highest resistance (ohms, Ω) setting and touch the probes to each end of the lead. Turn the dial to lower and lower settings until you get at least a two-digit number on the screen.

2. Leave the multimeter on the same setting and put the graphite next to a ruler. Touch one probe to one end of the sample and touch the other probe 1 cm away. Record the resistivity. Repeat for 2 cm, 3 cm, and 4 cm.

   Resistance 1 cm _____ Ω 2 cm _____ Ω 3 cm _____ Ω 4 cm _____ Ω

3. Look at the pattern of your data and circle which of the three equations are still possible.

   $R = \rho LA$ $R = \rho A/L$ $R = \rho L/A$

4. Repeat Step 2 with the 0.7 mm piece of graphite.

   Resistance 1 cm _____ Ω 2 cm _____ Ω 3 cm _____ Ω 4 cm _____ Ω

5. Now compare your results from Step 4 with your results from Step 2 and circle which equation must be correct.

   $R = \rho LA$ $R = \rho A/L$ $R = \rho L/A$

6. Use your data from the 0.7 mm diameter lead at 4 cm and the equation that you determined above to calculate the resistivity (ρ) for graphite. Use $A = \pi r^2$ to find the area and use 4 cm as the length.
7. Your teacher will give you the correct value for graphite's resistivity. Calculate your percentage error. Create your own data chart to enter the data.

Post-Lab Questions

1. Could you use this method to find the resistivity of metal wires?
2. Why do electricians have to use large wire if a lot of current will be flowing through it?
3. Why is it a bad idea to run large appliances through a very long extension cord?

Extension

Find samples of different kinds of metal rods and calculate their resistivity. You may use copper wire, aluminum or steel rods, and iron nails. Compare your answers to published data. You can look up the diameter of different gauges of wire on the internet.

LAB 63: SERIES RESISTORS

Students can easily see what series resistance means in this activity. Students are sometimes confused by the fact that all real resistors are the same size even when they have far different values. It is recommended that after this lab at home, students do a lab in class involving real resistors connected in series.

Data will vary depending on the pencil used and how dark the student colors. Students should be encouraged to color darkly and evenly every time.

Students will measure the resistance of resistors that they have drawn on paper with a graphite pencil. They will then connect two resistors in series and measure the resistance of the combination.

Topic: Resistors
Go to: *www.sciLINKS.org*
Code: THP38

Post-Lab Answers

1. Yes, the combination resistance is the sum of the individual resistances.
2. Yes, the longer they are, the higher the resistance.
3. So that when the gap is colored in, it doesn't add significantly to the resistance.
4. $R_{tot} = R_1 + R_2$

Note to teacher: Give students copies of page 274 to write on.

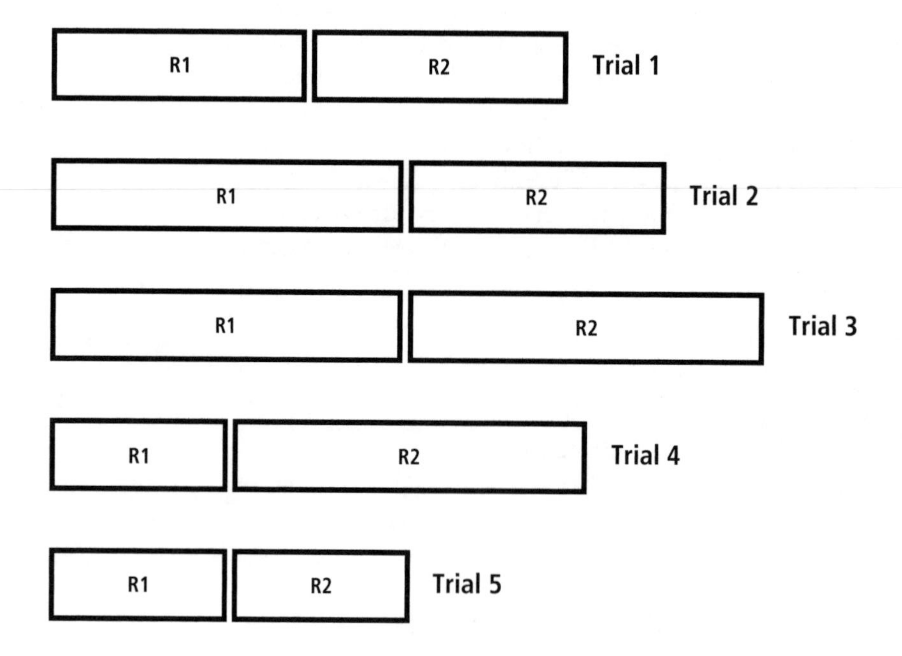
R1
R2
Trial 1
R1
R2
Trial 2
R1
R2
Trial 3
R1
R2
Trial 4
R1
R2
Trial 5

LAB 63: SERIES RESISTORS

QUESTION

How does resistance add up when the resistors are connected in series like in Figure 63.1, page 276?

SAFETY

Standard safety precautions apply.

MATERIALS

Graphite artist's pencil, multimeter, ruler

PROCEDURE

On the copy of Figure 63.2 that your teacher gives you, fill in the two boxes very dark, making sure that they do not connect. Measure their individual resistances and record the values in the data chart. Connect the two resistors by filling in the space between them and measure the combined resistance. Record the value in a data chart similar to the one on page 277 and repeat for Trials 2, 3, 4, and 5.

Figure 63.1

Representation of Resistors Connected in Series

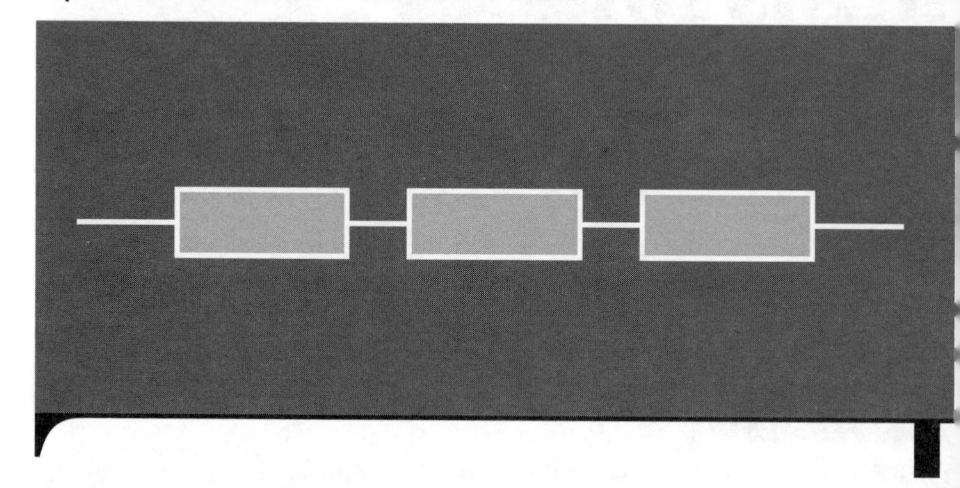

Figure 63.2

Resistors Aligned in Series

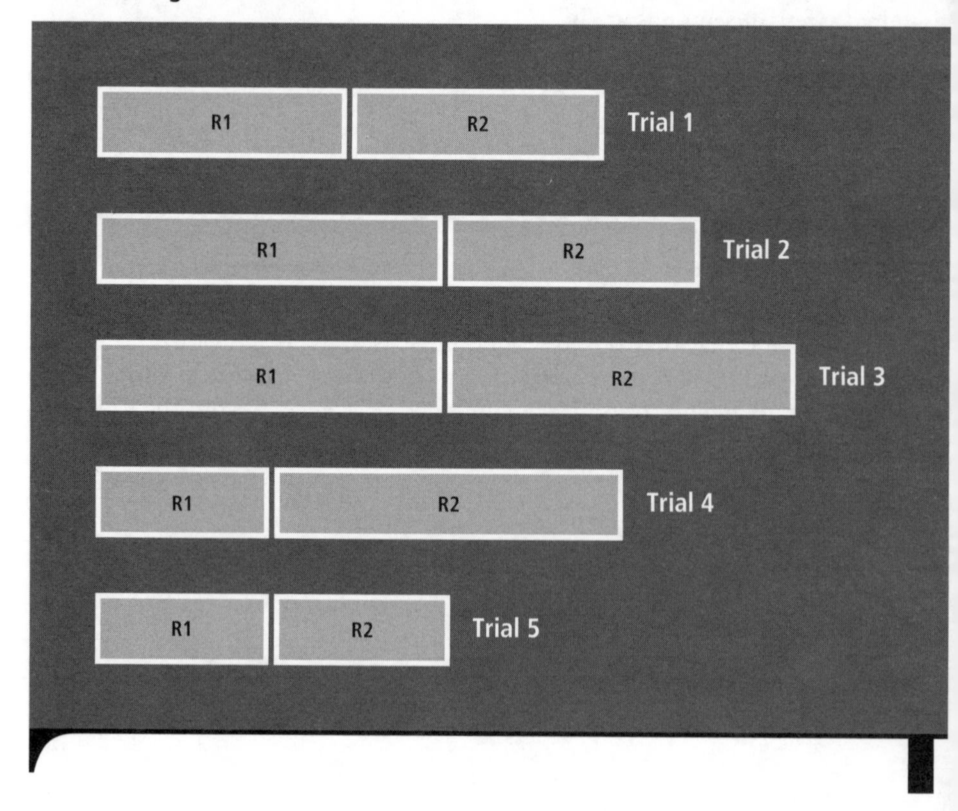

Data Chart

Trial	R1	R2	Combination
1			
2			
3			
4			
5			

Post-Lab Questions

1. Is there a mathematical connection between the individual resistances and the resistance of the combination? Can the connection be explained using the equation for resistivity?
2. Does the magnitude of the individual resistance affect the resistance of the combination? Explain your answer.
3. Why were the resistors placed so close together on the page?
4. Write a general equation for the sum of resistors connected in series.

Extension

Try placing three or more resistors in series and see if the same relationship exists. Use the equation for resistance and resistivity to explain this relationship.

LAB 64: PARALLEL RESISTORS

In this activity, it is important that students color very dark and completely. Otherwise, their results will be inconsistent. You may use standard pencils, but because they have waxes and fillers in their lead, the artist pencils work much better. The artist's pencils may be purchased from art supply stores for a couple of dollars and cut into three or four pieces. Then each of the smaller segments may be sharpened. Even a small piece of pencil will last for years.

The data in the chart will vary but should show the following patterns: In Trials 1, 2, and 5 the combination should be half the individual resistance. In Trials 3 and 4, the combination resistance should be less than the resistance of the small resistor.

Students will measure the resistance of resistors that they have drawn on paper with a graphite pencil. They will then connect two resistors in parallel and measure the resistance of the combination.

Post-Lab Answers

1. Sources of error include not coloring completely, impurities in the graphite, the extra resistance when the gaps are colored in, and touching the multimeter probes too far from the ends of the rectangle.
2. Yes, when the resistors are the same size, the combination is half the resistance of the individual resistor. In the other combinations, the total is always less than either of the individual resistances.
3. The resistances are not additive (1 + 1 = 2). The same size resistors result in a combination with half the resistance. The sum of any two resistors in parallel is less than either of the individual resistances. It doesn't matter which way the large and small resistors are arranged compared to each other.
4. Depending on the quality of the data and the students' mathematical reasoning ability, the teacher may have to assist with the equation. The equation is:

$$\frac{1}{R_{total}} = \frac{1}{R_1} + \frac{1}{R_2}$$

When teaching students to use the equation for parallel resistors, students will think that the equation can just be flipped over to $R_{tot} = R_1 + R_2$. Use some simple numbers to show them that this does not work; for example, $1/_1 = ½ + ½$ but $1 \neq 2 + 2$.

Note to teacher: Give students copies of the Figure 64.2 on page 281 to write on.

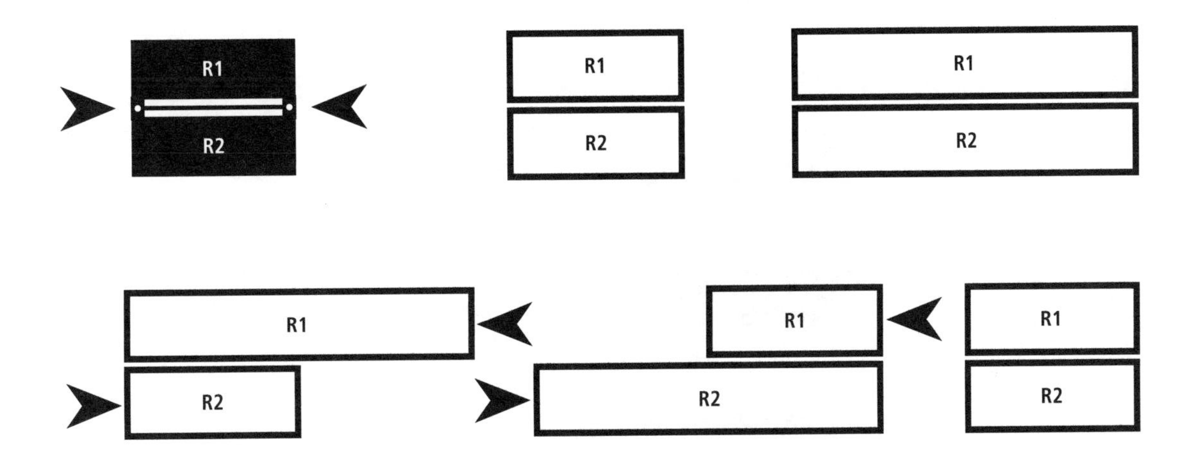
R1
R2
R1
R2
R1
R2
R1
R2
R1
R2
R1
R2

LAB 64: PARALLEL RESISTORS

QUESTION

How does resistance add up when the resistors are connected in parallel like in Figure 64.1?

SAFETY

Standard safety precautions apply.

MATERIALS

Graphite artist's pencil, multimeter, ruler

PROCEDURE

Your teacher will give you a copy of Figure 64.2 that you are allowed to color in. Fill in the two boxes very dark making sure not to let them connect. Measure their individual resistances and record the measurement in a data chart similar to the one on page 283. Connect the two resistors by filling in the space at the ends and measure the resistance from the white dots as in the example. Record this value in the data chart. Repeat this procedure for Trials 2, 3, 4, and 5.

Figure 64.1

Representation of Resistors Connected in Parallel

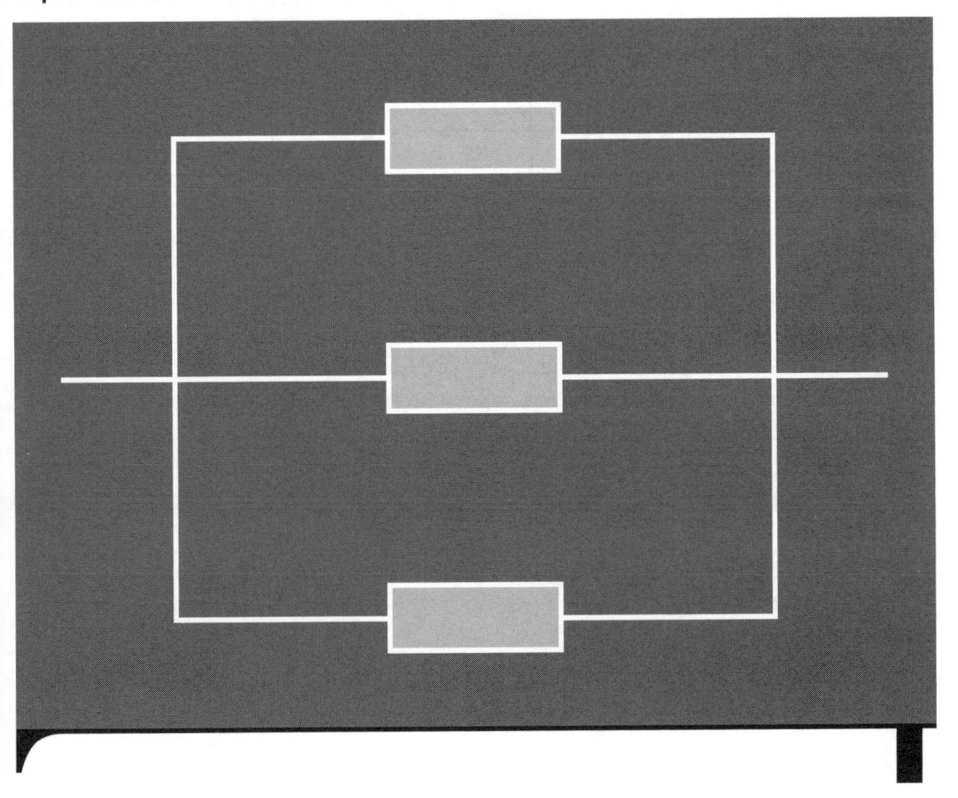

Figure 64.2

Resistors Aligned in Parallel

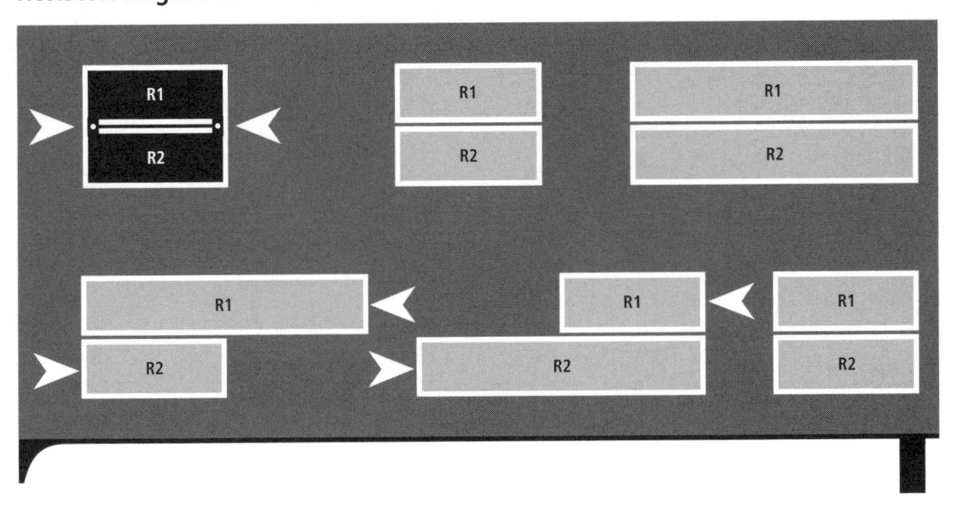

Data Chart

Trial	R1	R2	Combination
1			
2			
3			
4			
5			

Post-Lab Questions

1. What are three sources of error for this experiment?
2. Is there a mathematical connection between the individual resistances and the resistance of the combination? Describe the connection. Is there a special mathematical connection when the two resistors have the same value?
3. Write three general rules for the sum of resistors connected in parallel.
4. Write a general equation for the sum of resistors connected in parallel.

Extension

Try placing three or more resistors in parallel and see if the same relationship exists. Use the equation for resistance and resistivity to explain the relationship that you find.

LAB 65: SERIES/ PARALLEL BATTERIES

It is important for students to understand how resistors, capacitors, and batteries combine in series and parallel. The combination of batteries has a lot of practical applications in science competitions. This lab also reinforces how to use a voltmeter to test batteries, another practical skill. Students will learn that batteries connected in series add up their combined voltages. Batteries connected in parallel will add up their combined current output.

Post-Lab Answers

1. Series, +-+-+-+-, 6 V

2.

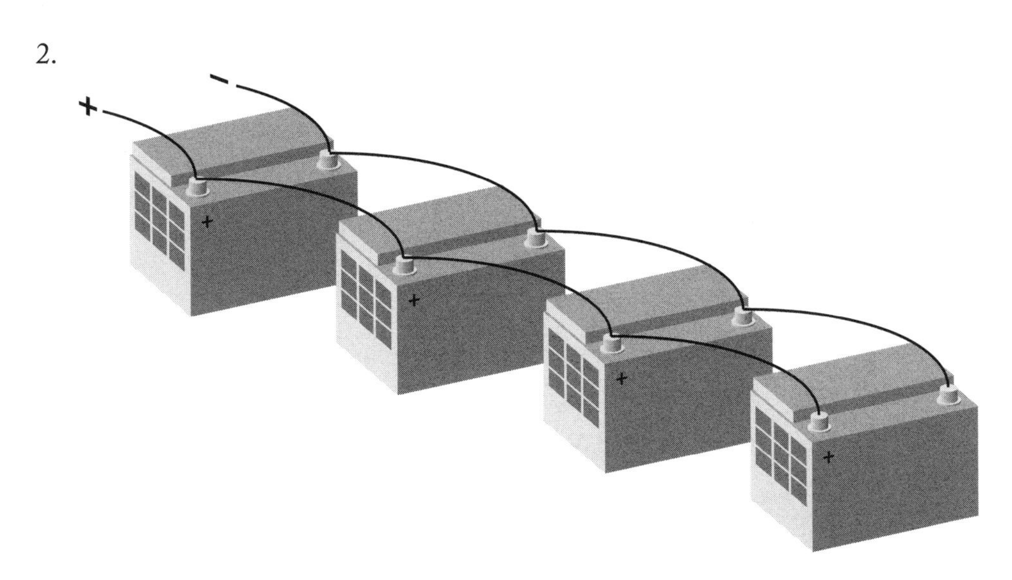

3. Series, 3 V

LAB 65: SERIES/ PARALLEL BATTERIES

QUESTION

How do batteries act when connected in series and parallel?

SAFETY

Only connect batteries as instructed. Connecting them any other way can result in overheating, chemical leakage, and possible injury.

MATERIALS

Two 9 V batteries, multimeter, two 6 in. wires

PROCEDURE

Batteries are sometimes connected in series and sometimes connected in parallel. One configuration gives a higher voltage and the other gives higher current. If a device needs 6 V, then four batteries of 1.5 V can be connected in the configuration that gives higher voltage. If you're putting a powerful stereo in your car that works at 12 V but needs high current, you would hook up 12 V batteries in the configuration that gives higher current.

1. Mark the batteries "1" and "2" so that you can tell them apart and use the multimeter to test the voltage of each one. Set the meter to a DC voltage setting of at least 10 V.

   Battery 1 ________ V   Battery 2 ________ V

2. Now use the wire to connect the batteries in series as shown in Figure 65.1 (p. 287). Be sure to pay close attention to which pips you are connecting. In this configuration, the batteries should go +, -, +, -. Now measure

the voltage from the first pip to the last. Be sure to set the meter high enough to handle up to 20 V in case the combination comes out to that much. What is the relationship between the individual voltages and the series voltage in this series connection?

Figure 65.1

Connecting the Batteries in Series

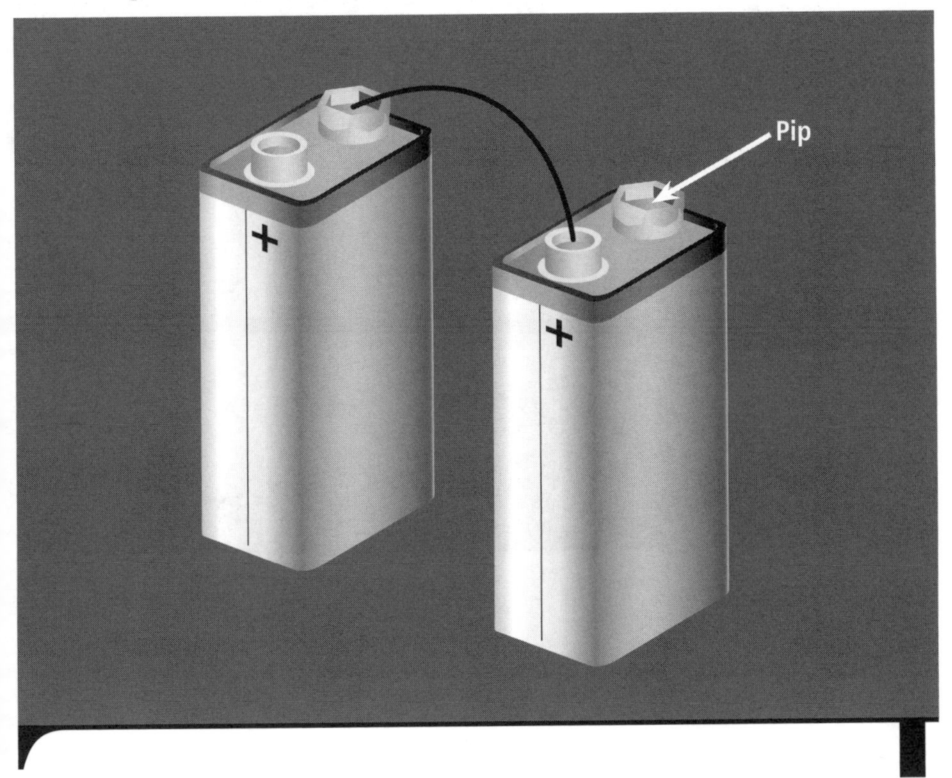

3. Finish the statement: When two batteries are connected in series, you connect the _____ pip to the _____ pip with a wire. The resulting voltage

   ____.

4. Connect the batteries in parallel as in Figure 65.2 (p. 288). Connect one positive terminal to the other positive terminal and one negative terminal to the other negative terminal. Now measure the voltage from one positive pip to one negative pip. Be sure to set the multimeter properly. What is the relationship between the individual voltages and the parallel combination voltage?
5. Finish the statement: When two batteries are connected in parallel, you connect the _____ pips together and the _____ pips together. The resulting voltage ________________________________.

Figure 65.2

Batteries Connected in Parallel

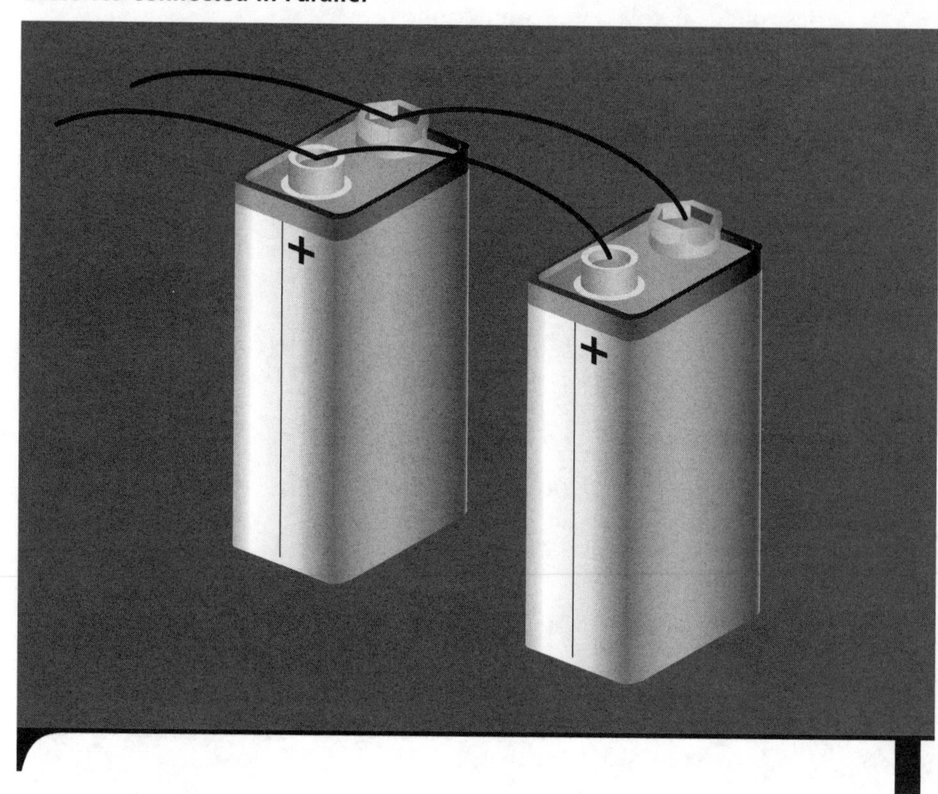

Post-Lab Questions

1. In a handheld video game, the batteries are connected as shown below. Are these batteries connected in series or parallel? If each battery is 1.5 V, what is the total voltage?

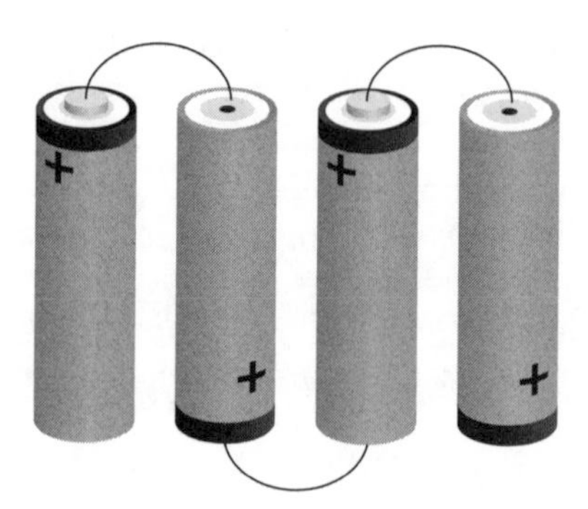

2. Finish the diagram showing how an electric car might have four 12 V batteries connected in order to end up with 12 V but high current capability. Show all the wire connections.

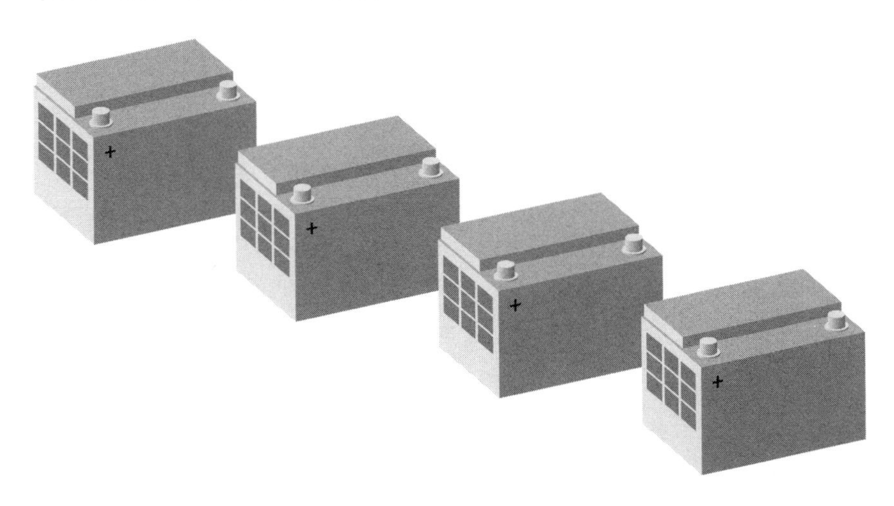

3. A flashlight has two 1.5 V batteries sitting on top of each other facing the same direction. Is this a series or parallel connection? What will the total voltage be?

INDEX